印迹 IMPRINT

深圳视界文化传播有限公司 编

Shenzhen Design Vision Cultural Dissemination Co., Ltd

新中式居住空间

LIVING IN NEO-CHINESE STYLE

中国林业出版社
China Forestry Publishing House

INHERITING ORIENTAL TREASURE, DEDUCING NEO-CHINESE CHARM
传承东方瑰宝 演绎新中式魅力

Having been in design industry for more than ten years and done many projects in different styles, for me, residence in Chinese style is my favorite. Perhaps because I flow Oriental blood and is very fond of enjoyable Chinese artistic conception of simplifying complicated materials. Traditional Oriental elements are expressed by modern design languages in the house; the combination of form and meaning creates traditional Chinese style into a new Chinese style which is suitable for modern people. This is a kind of cultural inheritance and also a kind of design innovation.

In recent years, after excessive pursuits of western grace and prosperity, the entire design industry begins to pay attention to natural and restrained lifestyle which Chinese people always advocate. "The room needs to be clean, elegant and concise; once there is something gorgeous, it seems that lady's attic is not suitable for the recluse to live in." Neo-Chinese style uses its beauty of artistic conception and cultural temperament to leave an impression on people. I think simple yet not dazzling, implicit, tranquil and graceful

Designs can comfort people's hearts. So in order to express Oriental culture, we should make good use of kind and deposited cultural symbols in traditional style, transform them into contemporary forms by our skills and interpret unique Oriental tranquility on spiritual and cultural levels.

Neo-Chinese style is not simple piles of retro elements but three aesthetic tastes that we understand Chinese tradition in modern vision, which are the beauty of justice and peace, the fun of elegance and wideness and the legend of folk art. The combination of classic and concise methods outlines endless images and coveys purer beauty of Chinese artistic conception. When designing Downing Mansion, I choose plain and clean green to bring tranquil and meaningful flavors; repeated use of colors makes the space harmonious and unified. Firmness in general parts and softness in details make the dignified and elegant atmosphere present cultural connotations.

Mr. Liang Sicheng once said: "If the old Oriental city completely loses its artistic features on architecture, it is painful both on cultural performance and appreciation." As designers, we take on the responsibility of inheriting Chinese culture and should think and innovate constantly to make Chinese living space lead the trend of time and carry it forward at the same time bring people better life yearnings.

RUI NING
GND DESIGN LIMITED N+

从事设计行业十几年，做过大大小小许多不同风格的项目，于我而言，中式风格的住宅，却是心头之爱。也许是因为身上流淌着东方人的血液，也许是酷爱中式删繁就简、写意于心的意境。传统东方元素在室内用现代设计语言表达出来，形和意的结合，将传统中式风格打造成适合现代人居住的新中式风格，这是一种文化上的继承，也是一种设计上的创新。

近年在过度追逐西方的雍容繁华后，整个设计界开始关注到国人始终崇尚自然、含蓄节制的生活方式，"室中清洁雅素，一涉绚丽，便如闺阁中，非幽人眠云梦月所宜"，新中式以其意境之美以及其缓缓流淌的文化气质打动着人们。而我认为朴拙而不耀眼，含蓄如处子般静谧婉约的设计，置身其中就能让心灵得到慰藉。因此对东方文化的表达，要善用传统中良善、沉淀的文化符号，再用我们的技艺转化为当下的形式，从精神、文化层面来诠释东方特有的静谧。

新中式不是复古元素的简单堆砌，而是以现代的眼光理解中国传统的三种审美趣味，中正平和之美、文雅野逸之趣及民间艺术之奇。用经典与简洁相结合的方式勾勒出无限意象，将中式的意境之美表达的更加纯粹。在我设计唐宁府时，选择使用素净的绿色带出宁静隽永的气息，色彩重复使用让各空间和谐统一。于大处见刚，细部现柔，使庄重典雅的氛围中流露出文化底蕴。

梁思成先生曾经说过："一个东方老国的城市，在建筑上，如果完全失掉自己的艺术特性，在文化表现及观瞻方面都是大可痛心的。"作为设计师，我们也肩负着中华文化传承的责任，在带给人们美好生活向往的同时，应不断思索和创新，使中式居住空间引领时代潮流并得以发扬光大。

宁睿
香港杰地设计集团 深圳市恩嘉陈设艺术设计有限公司

CONTENTS / 目录

印迹 IMPRINT

006 希望栖 白鹭城
Habitat of Hope, City of Egret

016 大艺术家·繁秋
The Artist, Flourishing Autumn

028 若竹
The Same as Bamboo

036 箫鼓瑟瑟 古风犹存
Rustles of Flute and Drum with Ancient Charms

044 邂逅文化的交融
Encountering Cultural Integration

054 偶然入林处 谈笑便无期
Endless Chatting in the Forest by Accident

070 景以致远
Far-Reaching Scenery

080 小隐隐于野 大隐隐于市
Retreating into the Noisiest Fair is the Greatest Hermit

092 青墙瓦黛
Green Wall and Dark Tile

106 半窗疏影恋黄昏
Half Window Flickering the Fading Shadow in the Twilight

118　藏形露意
Hiding Form and Exposing Meaning

134　素雅空间
Simple and Elegant Space

140　山水之间 诗意栖居
Poetic Residence Between Mountains and Waters

150　雅藏
Elegant Collection

160　源于内心 写意山水
Enjoyable Landscape Inspired by Inner Heart

168　水墨山行
Ink Painting Mountain Trip

178　相遇时光
Meeting with Time

186　留·白
White Space

194　云裳荷影 落红拂柳
Shadows of Cloud and Lotus, Flowers Falling Through the Willow

204　清风朗月照人眠
Sleep with Cool Breeze and Bright Moon

214　闲来话尽花满庭
Spare Time at Courtyard Full of Flowers

230　临风听蝉 青墨雅涵
Listen to Cicadas with the Wind, Feel Cyan Ink and Elegant Connotation

240　黛绿年华梦似锦
Splendid Dreams in Dark Green Times

250　幽深古色
Deep and Quiet Ancient Color

266　泊墨之境
Ink State

278　忆·相随
Memory and Accompanying

288　隐庐
Secluded Residence

294　一念姑苏 水云间
One Thought of Suzhou, Water and Cloud Residence

302　端庄素影
Dignified and Plain Shadow

308　繁华尽处 归于璞素
Simplicity After Gorgeous Prosperity

IMPRINT 印迹

郭子伟
姜雪
ZIWEI GUO
CHER

希望栖 白鹭城
Habitat of Hope, City of Egret

项目名称	新希望白鹭城F09地块样板房
硬装设计公司	朴悦设计
软装设计公司	深圳市寐卡国际
参与设计	彭柯麟、刘国琴
项目地点	浙江温州
项目面积	230 m²
主要材料	水曲柳、富春色、金箔、贝母漆、爵士白大理石、高档皮料布料等

DESIGN CONCEPT | 设计理念

As neo-Chinese style, this project uses extremely simple modern technique to run through and present traditional Oriental spirits and elements. Combining with concise and sedate space setting, this project portrays Chinese lines in modern aesthetic framework. On materials, the designer manly uses velvet and wood to present its natural Oriental texture and uses proper trail leather and metal to present details. The entire space is based on the intermediate tone with clear layering in it, as if the use of Chinese ink painting technique "five divisions of ink color" which are "parched, dense, heavy, light and clear" with colors as embellishments. The furnishings are in line with slap-up, exalted and comfortable concepts; the designer uses cross-boundary technique to endow the whole space with elegant and high-end temperament, to complete its functions and practicability and to convey an exquisite and elegant life aesthetics.

　　本项目设定为新中式风格,将传统的东方精神与元素以致简的现代手法予以贯穿、呈现。结合简洁稳重的空间设定,以中式线条描绘于现代审美的框架内。材质上多用丝缎、木质表现其自然般的东方肌理,并适度试用皮质及金属表现细节。整个空间以中间色调打底,但其中却划分出清晰的层次。好比中国水墨技法"墨色五分"中"焦、浓、重、淡、清"的运用,最后再加以色彩作为点缀。配饰上本着高档、尊贵、舒适的理念,采用跨界的手法,赋予整个空间优雅高端的气质,完善其功能性和实用性,传达一种精致典雅的生活美学。

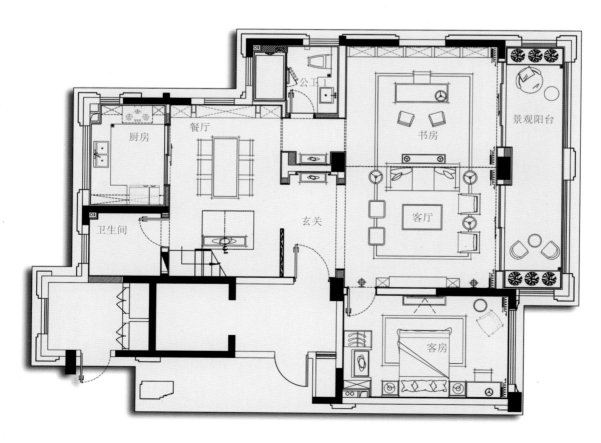

印迹 IMPRINT

刘卫军
DANFU LAU

大艺术家·繁秋
The Artist, Flourishing Autumn

项目名称 ｜ 金茂·佛山绿岛湖项目示范区别墅样板房250户型
设计公司 ｜ PINKI DESIGN美国IARI刘卫军设计事务所
陈设配饰 ｜ THE ARTIST大艺术家软装设计与实现
项目设计 ｜ 陈春龙、黎俊浩
艺术主创 ｜ 李莎莉
软装统筹 ｜ 张慧超
项目地点 ｜ 广东佛山
项目面积 ｜ 530 m²
项目摄影 ｜ 曾朗

DESIGN CONCEPT ｜ 设计理念

With neo-Chinese style and modern technique as the main line of thinking and collocating with Chinese Zen artistic conception, this project has enthusiastic and bold cultural temperament with spiritual sense of belonging. The expression of cultural connotation and interpretation of fashion are deduced incisively and vividly, which foils cultural flavors of the owner and brings cultural and noble life quality experience to the owner.

以新中式风格与现代手法为思维主线，配合中式的禅意意境，使本案在文化气质上热情奔放又不失心灵的归宿感。文化底蕴的表达与时尚的诠释也演绎得淋漓尽致，从而衬托出主人的文化气息，给主人带来文化与高贵的生活品质体验。

The space decoration uses concise and hale straight lines, chooses typical red furniture and modeling furnishings and collocates them with Chinese style. The application of line decoration in the space not only reflects modern people's living requirements of pursuing simple life, but also caters to restrained and plain design style of Chinese furniture, which makes Chinese style more practical and full of modern sense.

空间装饰相对采用简洁、硬朗的直线条，选择具有典型红色的家具与造型装饰，搭配中式风格来使用。直线装饰在空间中的使用，不仅反映出现代人追求简单生活的居住要求，更迎合了中式家具追求内敛、质朴的设计风格，使新中式更加实用，更富现代感。

负一层立面图

客厅立面图

Rich decorative detail is the sublimation of the traditional Chinese style. The furnishings can reflect tastes of the owner and enrich cultural connotation of the space, which is also inherited and reflected in neo-Chinese style. Chinese ancient clothes of the sofa background in the living room is presented in the form of art pendant and manifests cultural accomplishment and appreciation level of the owner. It collocates with the white marble, which makes a new sublimation. Hale metal strips and plain wood veneer perfectly deduce a dynamic and static, hard and soft combination.

丰富的装饰细节是传统中式设计的升华，其中饰品可以体现主人品位，丰富空间的文化底蕴，这点在新中式上同样有所继承和体现；会客厅沙发背景的中国古代服饰以艺术挂件的形式呈现，彰显出主人的文化素养与欣赏水平，搭配白色大理石使高贵有一个新的升华；硬朗的金属条与朴素的木饰面，呈现出动与静、刚与柔的完美演绎。

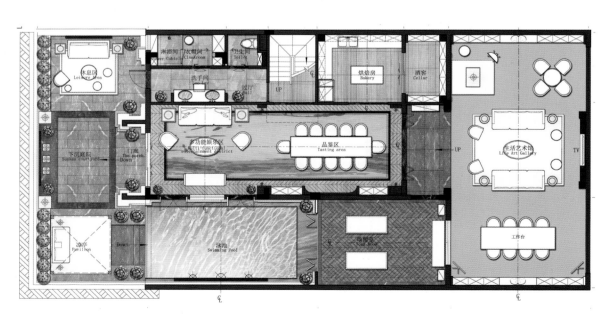

As the leisure and recreational space, the negative layer has functions of reception, gathering within family members, leisure and recreation. The space starts from the garage to inner tasting area, multi-functional recreational area and baking room. The corridors form a transparent axis which divides different functional spaces. The garage can become a space to taste art for friends and families when needed. The swimming pool and multi-functional recreational area interact; the liquidity of the water scene wall makes the space full of vitality with a dynamic and static combination. The way of enclosing of the multi-functional recreational area is relaxing and comfortable, which creates a good communication space.

负一层为休闲娱乐空间，承当着对外接待及内部家庭成员的集会、休闲与娱乐等功能。空间由车库往内层层递进于品鉴区、多功能娱乐区、烘焙房，过道形成一条通透的中轴线沿着中轴线分了各个功能空间，车库在需要的时候可以变成亲朋好友品鉴艺术的空间，泳池与多功能娱乐区的互动，水景墙的流动性让空间富有生气，动中有静，静中有动，而多功能娱乐区的围合方式，轻松舒适，给空间营造了一个良好的交流场所。

The first floor is mainly for family members and guests and is set with art gallery, living room, dining room and Chinese kitchen. The living room has open view after integrating the space layout; its rigorous axis symmetric relation presents spaciousness and magnificence of the space. The second floor is sleeping area. The combination of children's room, study and cloakroom makes the space function superior. The elder's room has an independent cloakroom, making the space nobler. The recreational balcony is an extension of the interior space. The virescence and featured landscape endow the balcony with a feelings as if in the forest. The third floor is master bedroom activity space. The perfect combination of sleeping area and reading area gives the owner relaxing and pleasure reading time. Luxurious master bedroom brings the owner custom-made extraordinary experience and enjoyment.

一层以家庭成员及会客为主，设有艺术长廊、会客厅、餐厅、中厨，整合空间布局形成宽敞的会客厅，开阔的视野，严谨的轴线对称关系，展示了空间的开阔大气。二层为睡眠区，小孩子房与书房衣帽间的结合，使空间的功能性更为升级；长辈房拥有独立的衣帽间，让空间更为高贵；休闲露台是室内空间的延伸，通过绿化和景观的小品设置，将露台营造出如同置身于森林的感觉。三层为主卧活动空间，睡眠区与阅读区的完美结合，让业主享受轻松惬意的阅读时光，奢华的主卧空间带给业主私人定制式的超凡体验与享受。

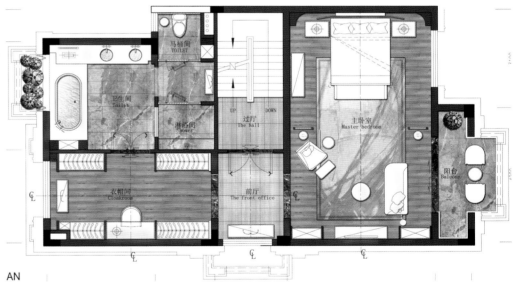

AN

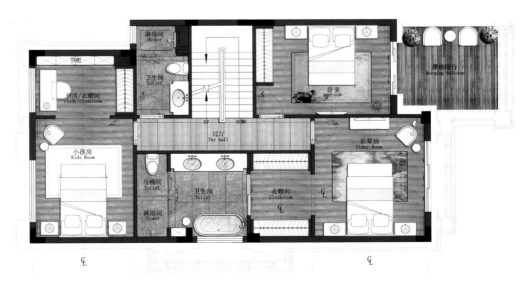

主卧立面图

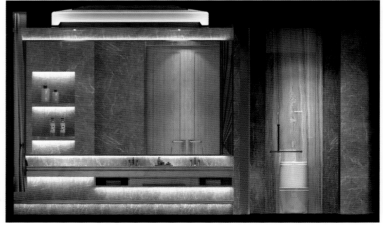

主卫立面图

印迹 IMPRINT

陈嘉君
邓丽司
ALAN CHAN
ALICE DENG

若竹 The Same as Bamboo

项目名称 ｜ 佛山保利云上
室内设计 ｜ C&C壹挚设计
软装设计 ｜ C&C壹挚设计
参与设计 ｜ 陈秋安
项目地点 ｜ 广东佛山
项目面积 ｜ 280 m²
摄 影 师 ｜ 冯健、程小乐

DESIGN CONCEPT ｜ 设计理念

Chinese people's love to Chinese style is permeated in bones. Culture cannot get rid of historic foundation; how to transplant its integrity and influence into nowadays to make it a part of design and find its root and essence is what we need to think constantly and want to realize in design process.

　　中国人对于中式情怀的喜爱应该是渗透在骨子里的。文化始终不能脱离历史的根基，如何将它的完整性和影响移植到今天，让它成为设计中的一部分，找到根基与精髓所在是我们在设计的过程中不断思考和想要实现的。

This time we use a lot of bamboo and rattan elements. Droplights with changing modeling try to create a beautiful and ethereal feeling; large area of white space of the abstract ink painting on the wall depicts the sense of scale and layering of the space, which is kind of like the meaning that "the white clouds intentionally pass through the bamboo with clear tweets". Sunshine comes in from the outside world in the morning; the light and shadow of cloud sways with wind and light, static or dynamic; when meeting with order bamboos, the lights are divided into many pieces, scattering in the space; the residents can experience the flowing light and shadow and enjoy different sceneries every day.

这次我们运用大比例的竹子和藤编的元素，造型多变的吊灯试图在这个空间营造隽秀飘逸的感觉，墙上大面积留白的抽象水墨画，刻画出空间的尺度感和层次感，倒是有几分"白云如有意，穿竹伴清吟"的意味。阳光随一日晨昏从外面的世界渗入进来，云的光影随着微风、光照溶溶曳曳，或静或动，遇上疏密有致的竹子，光线便被分成了许多碎片，散落在空间里面，居住者能够体验其间的光影流动，享受每日不一样的景色。

Going to the Orient and tracing to the source of China. Time is changing, so are human's aesthetic principles and pursuits. Rooting in the essence of traditional Chinese culture, we try to use Oriental blood from Chinese to reshape our definition of beauty, push away all established conformations to integrate Oriental and modern connotations. We hope that what the space presents is Chinese feeling full of modern sense and what it stresses is Oriental implication.

至境东方，溯源中国。时代在变，人的审美原则和追求也在变。扎根于传统中式文化的精髓，我们尝试用中国人流淌的东方血液重塑我们对美的定义，推开所有既定的符合，把东方和现代的意蕴糅合起来，我们希望在空间呈现的是富有现代感的中式情怀，讲究的是东方的意韵。

We try to use modern people's aesthetic needs to create things full of Chinese traditional charms and endow the space design with a new Oriental style to continue the context of Oriental traditional art. Oriental spirit and Western form are mixed together and are used throughout the whole space by a manner of "mind changing with different environments", which makes the residence full of flexibility as if in the landscape where tranquil and ethereal clouds among the mountains come here with the wind.

我们试图以现代人的审美需求来打造富有中国传统韵味的事物,将空间设计注入一股崭新的东方风格,使东方传统艺术的脉络得以蔓延下去。东方精神和西方形态交织结合起来,以一种"心随万境转"的精气神贯穿整个空间,使居室充满灵动性,仿佛置身山水,恬淡空灵的山间云雾,随风而动来到这里。

印迹 IMPRINT

李健
JACKY

箫鼓瑟瑟 古风犹存
Rustles of Flute and Drum with Ancient Charms

项目名称｜华夏幸福基业·廊坊固安孔雀城·合院别墅样板房
设计公司｜PCD品仓设计
参与设计｜梁祖敏、殷雾
项目地点｜河北廊坊
摄影师｜三像摄
主要材料｜蓝金沙、实木染色、皮雕、丝麻墙纸、古铜不锈钢、拼花木地板、浑水漆等

DESIGN CONCEPT ｜ 设计理念

Our design has never been a piece of display item with fierce emotions but a work of art with everlasting appeals.

The No.66 courtyard in our eyes is supposed to have an inheritance of attitude, be based on the long Oriental culture to conduct conscious development and remodeling and endow itself with a temperament of a capital mansion to skillfully hide in the city texture. Neither abrupt nor catering, it fully shows the confidence from Orient.

　　我们的设计从来不是一件情绪激烈的展览品，而是拥有持久感染力的艺术品。

　　在我们眼里的66号合院，本应具备一种传承的姿态，以悠久的东方文化为基底，进行有意识的发扬和重塑，令其具有一种京城大宅的气质，巧妙隐于这城市肌理之中。不突兀亦不迎合，并充分展现出来自东方的自信。

With time and space as the bond, Chinese No.66 courtyard integrates Oriental calmness and spirituality, exquisitely depicts warmth and tranquility by time traces, creates contemporary humanistic elegance by modern Oriental languages and explores uncopiable humanistic connotations in this piece of land.

At the entrance porch, opening the door, it comes straight to the point. The flowing marble texture seems to indicate that "a river is embracing the fields like an emerald girdle; the two hills on both sides are like a gate open, streaming green shade in flow." Piles of mountain shadows make people look happy and enjoyable in the refreshed atmosphere. Entering the living room, the light long sofa with high back makes a contrast with the tripping carpet. The warm tone makes people warm and comfortable; the dialogue between the owner and the guests reaches the situation that "sharing views and tastes they drink heartily to each other, while their horses are tethered to willows by tall buildings." Small delicate furnishing articles, blue and yellow contrasts, pearl fish skin tray and platinum artworks present noble temperaments and set off the entire space. The dining room continues the warm tone in the living room. It is worth mentioning that hand embroidery of the dining chair back is created by female embroiders of Suzhou. Warm lights pass through gold copper acrylic texture and set off on the gilded tableware, creating a magnificent and dignified atmosphere.

华夏66合院以时空为纽带，融入东方沉稳灵性，用时光的痕迹细腻刻画温润与澄静，运用摩登东方语言营造当代人文高致，探寻这片天地不可复制的人文意蕴。

入户玄关，开门见山。大理石流动着的纹理，仿似在诉说着"一水护田将绿绕，两山排闼送青来。"山影层峦叠翠，神清气爽间喜上心头，乐上眉梢。步入客厅，一座高背浅色长沙发，与地毯的跳脱形成对比。暖色调的选择让人顿觉温馨舒适，主客谈笑间，岂不"相逢意气为君饮，系马高楼垂柳边。"小巧精致的摆件，蓝黄对比色的凸显，珍珠鱼皮托盘，铂金艺术品，彰显着尊贵气质，与整体空间相得益彰。餐厅延仗客厅的暖色调，值得一提的是，餐椅椅背的手工刺绣，来自苏州绣娘一针一线织就而成，十分重工。暖色的灯光透过金铜亚克力质面，映衬在描金漆餐具之上，粼粼微光，营造出大气、贵重的氛围。

"In a winter night, the guest comes and is treated with tea instead of wine; the houseboy is asked to boil tea; the stove flames become red; the water is boiling in the pot; and the house becomes warm." Tea culture has a long history. "The fragrance which is from neither grasses nor flowers" needs some bosom friends to taste and perceive. Handmade pot twined with old vines is filled with boiling water; the steam is rolling up along with tea fragrance; the swaying leaves in ups and downs quietly remove the fickleness; one can feel the inner cleanness and brightness, and all is quiet. Treating the guests with tea is social etiquette of the ancients, brings a quiet, beautiful and meaningful artistic conception for the friends and is regarded as an elegant thing. Therefore, there is a tea room in the first floor, which offers a better place for a chat and a cold night.

"寒夜客来茶代酒，竹炉汤沸火初红"。茶文化源远流长，那"非叶非花自是香"也需约上三两知音去品、去悟。老藤缠绕手工打造的壶，盛着烧开的水，蒸汽携着茶香袅袅上升，沉浮间摇曳的芽叶，也静静地磨去了浮躁，感心中之清明，唯万籁皆寂静。以茶待客，乃古代人情交际的礼节，它为友人之间带来一种清幽隽永的意境，更被视为风雅之事。故此，一层设禅茶室，供长日清谈、寒宵兀坐。

Downstairs, jagged metal ball droplights pour down; the fancy texture seems to be exquisite landscape paintings; the hazy lights along with the landscape grains are like the thousand cultural river, pouring onto a group of sculpture *Home* below and forming a delightful contrast. *Home* depicts a picture that a man and a horse march forward in the distant mountain. The front road is flat and fluent; when the horse arrives, the goal is realized. The hometown is near at hand; the long way is quiet with a grand ambition that "an old war-horse may be stabled, yet it still longs to gallop a thousand li".

The recreational area at the bottom of the stairs belongs to the host of the villa. The carpet as if a huge piece of landscape painting sets a tone for the entire space. As a social space, there collects his precious memories from his travels.

"Harmonious and different, everything is in the proper place." In the second floor, the gilded painting sets a tranquil and restrained tone for the master bedroom. The space furnishings with peaceful temperament develop and gain the continuation and innovation in the inheritance after precipitation of time. The situation that "my house is free from worldly moil or gloom, while ease and quiet permeate my private room" by Tao Yuanming makes your heart break through the barriers to return to the self and enjoy the wonderful life. Collections and feelings integrate inseparably.

顺着楼梯循级而下，参差不齐的金属球形吊灯灯影如泄，幻化的肌理，犹如一幅幅精美山水画，顺着山水的纹路，透出朦朦胧胧的光如数千年的文化长河，倾泻于底下的一组雕塑《归宿》之上，与他相映成趣。《归宿》描摹出了一人一马远山上行进的画面，前方的路平坦顺畅，马到即成功。家乡已近在咫尺，迢迢的路啊，虽自在清静，却也不乏"老骥伏枥，志在千里"的豪情壮志。

位于楼梯底部的休闲区，专属于别墅的男主人，地毯仿若一幅巨大的山水画为整个空间的态度定调，作为社交空间，这里更私藏了他游历四方遇见的珍贵记忆。

"和而不同，各得其所"。来到二楼，映入眼帘的金箔画，为静气内敛的主卧渲染基调。空间饰品取自岁月静好之气质，历经时间沉淀，在传承中发展，在传承中得到延续与创新。像陶渊明那般"户庭无尘杂，虚室有余闲"，让自己的心灵冲破藩篱，回归自我，享受精彩、享受人生。置物与情怀，相融相和，难舍难分。

Children's rooms take this suggestion. The real "beauty of Chinese style" borrows the power of the field itself and rich colorful furnishings to endow the space with fresh vitality in cultural acceptance, in influential aesthetic consciousness and in the world of light and shadow. Girl's room with pink fantasy and boy's room with energy and vitality are injected with Chinese Zen, manifesting the unity of artistic expression and artistic conception.

儿童房以此为谏，那么真正的"中式之美"，也只有在文化的接纳中，在耳濡目染的美感意识中，在光与影的世界，借助场域本身的力量和丰富的色彩饰品装饰摆件，为空间关注鲜活的生命力。女孩房的粉色梦幻空间，男孩房的活力动感，两者融入中式的禅意空间，无不彰显其艺术的表现与意境的合一。

印迹
IMPRINT

杜文彪
BILL DU

邂逅文化的交融
Encountering Cultural Integration

项目名称 | 广州杜文彪装饰设计有限公司
参与设计 | 谭莎、胡莎莎、张惠秀
项目地点 | 北京
项目面积 | 685 m²
摄 影 师 | Bill
主要材料 | 木饰面、天然石材、金属、丝布硬包、实木复合地板等

DESIGN CONCEPT | 设计理念

The location of this project surrounded by Wenyu River and "Fangshi Canal" is a place which is "not far away from prosperity and close to nature". The architecture is created by "garden" and "yard" which interpret core of Chinese traditional architecture by modern technique.

Tranquility and freedom are largest luxuries in the world and cultural collision can also inspire new senses. The collision of Oriental and Western cultures reflects in the conflict and opposition of cultures. Aesthetics of ancient people and lifestyles of modern people are integrated into one.

本案所在地由温榆河和"方世渠"围合而成的宗地是一处"离繁华不远，离自然更近"的场所。建筑以"园"与"院"进行打造，两者之间则以现代的手法来诠释中国传统建筑的内核。

安静和自由是这个世界最大的奢侈，文化的碰撞也能激发新的感官。东西方两种文化的碰撞体现在文化的冲突与对立上，古人的美学，今人的生活方式，合二为一。

Chinese phoenix uses spun silk wall cloth hand embroidery to pave, appears in front of people in a more modern and lightsome form and creates a natural, warm and noble atmosphere. From outside to inside, fluent lines collide and intersect with each other, rising slowly and releasing tranquility, which contributes to "home reunion".

The overall space designs increase profound connotations by diluent delicacy with gentle pace. The materials and patterns of furniture abandon complicity and exaggeration, whose neutral colors create an endless comfort in proportion, emotion and story.

中式凤凰使用绢丝墙布手工刺绣，层层叠叠铺就，以更现代、轻盈的形式出现在世人面前，营造自然和暖不失贵气的氛围。由外到内，流畅的线条相互碰撞交汇，缓缓上升，释放静谧，促成"归家团聚"。

整体空间设计，以冲淡的精致叠加深厚的底蕴，轻柔踱步。在家具材质和款式方面摒弃繁复浮夸，中性的色调在比例、情绪和故事间平衡出了无限的舒适。

The great painter Mr. Qi Baishi thinks the painting charm lies between similarity and differentia. Ink painting foils the magnificence and flexibility of the space, adds a humanistic and design temperament and integrates humanity, art and life into a whole. The combination of Chinese and Western, enthusiastic and vivid colors and strong contrast of ink and color present Western bold colors and Oriental profound charm quietly. Wandering here, the mixed feeling of unique Chinese and Western languages arises spontaneously.

画家齐白石先生认为作画妙在似与不似之间。水墨画烘托出空间的大气和灵动感，增添了一份人文与设计气息，将人文、艺术和生活融为一体；中西交织，热烈明快的色彩，墨与色的强烈对比，潺潺流水中尽显西方奔放色彩和东方浓厚神韵。漫步于此，独特的中西语言的交织感油然而生。

印迹 IMPRINT

INGALLERY™

偶然入林处 谈笑便无期
Endless Chatting in the Forest by Accident

项目名称 | 绿城·南京桃花源样板间
项目地点 | 江苏南京
项目面积 | 550 m²
摄 影 师 | ingallery™
主要材料 | 木质、藤、地砖等

DESIGN CONCEPT | 设计理念

As early as 1500 years ago, there is a small town in Tangshan piedmont in Nanjing, Jiangsu, which is called "holy land" by the royal. In the forest, there are many courtyards and houses. After Ming and Qing Dynasties and the Republic of China, leisurely enjoying the landscape, focusing on gardens and lingering on hot spring, people who live here know elegance and enjoyment most.

You don't traverse back to a house of an ancient officer. This is Greentown Nanjing Peach Garden. Everywhere reveals the beauty of Chinese courtyard. With rockery, fantastic rock peaks and woods, the movement of steps and shift of scenes constitute a different world. It is worth mentioning that Mr. Xue Lingen as " the inheritor of Xiangshan National Intangible Cultural Heritage" uses skills for more than 600 years to create this Chinese courtyard villa; being in this house, one can feel as if in Suzhou Garden.

早在1500年前，在江苏南京汤山山麓有一小镇，被皇家封为"圣汤之地"。山野之中，隐藏着处处庭院和户户人家。经明清，越民国，悠游于山水，寄情于园林，流连于温泉，有着一群最懂风雅和享受的人。

你没有穿越回到古代某大官的院落宅邸，这里是绿城·南京桃花源，这里处处透露出中式合院的美。假山、怪石、林木，移步换景，构成了另一个世界。而这不得不提作为"香山帮非遗传人"的薛林根老师，其用沿袭600多年的技艺，打造了粉墙黛瓦的中式合院别墅，让置身绿城·南京桃花源中，却有一种来到苏州园林的感受。

Here is far away from the city hustle and bustle with pure air in the mountain and sounds of bird. Walking through the yard, you can be attracted by the small rest place, listen to the sound of rain in the corridor and taste tea in the yard to enjoy Zen in life which is as if the faint scents and aftertastes of bitter tea.

Outside the French window of the living room, you can see soft and tranquil water scene in the yard, which is "borrowing scenes by leaning on the window". The symmetric set-up of furniture and lamps endows the space with sense of ritual. Wood tea table and vine sofa are relaxing and comfortable; every detail of the ancient folding chair from Italy reveals high pursuit of life. Natural landscape grain stone screen and the opposite ancient painting echo with Tangshan landscape, which makes it more tranquil and reveals the beauty of earth and heaven. The Chinese dining room is a place for the family to reunite and eat with peaceful atmosphere. Closer to the sedate and magnificent bedroom, the dark blue and gold enrich the space layering. The 16-year-old girl's room has elegant texture of a mansion where the vigorous bright yellow jumps freely. The boy's room is simple and handsome; cute animal furnishings create surprises; the plain geometric carpet is just to the point. The corner for tea is concise and clear; what you see and what you want are full of endless Zen. As main decorative patterns, the landscape grains are refined and reorganized, which are like clouds and waters in classical and modern styles and like mountains and hills in Chinese and Western styles.

　　这里远离都市喧嚣，只有山中纯净的空气和虫鸟鸣叫。走过庭院，被其中的小坐之处吸引，可在廊下听雨，也可在院中品茗，享受生活中的禅意，就像苦茶中的清香、回味无穷。

　　起居室里的落地窗外，就能看到庭院柔和幽静的水景，"倚窗借景"便是如此。对称的家具及灯具摆放，使空间更具仪式感。木质茶几与藤艺沙发放松又舒适，来自意大利的品质交椅，每个细节都透着对生活的高度追求。从天然山水纹石材屏风，到与之相对的古画，都与汤山治山水遥相呼应，这里愈发静谧，天地灵秀立显。中餐厅，一家人团圆吃饭的地方，充满祥和之气。沉稳大气的卧室，走近才发现，点缀其中的深蓝、暗金，空间层次极为丰富。十六岁少女的闺房，既有大宅的优雅质感，又多了几分青春活力的明黄色，尽情地跳跃着。男孩房简单帅气，可爱的小动物饰品创造惊喜火花，几何素色的地毯再合适不过。用来品茶的一角则简约明了，目之所及与心之所想都是无边无际的禅意。选择山水纹作为主要的装饰图案，经过提炼和重组的山水纹，似云似水、亦古亦今，似山似峦、中西并举。

The designs integrate pure Chinese garden yard residential culture, use elegant and implicit black, white and gray as tone of the space and collocate with brown walnut veneer, which is sedate, low-key and restrained.

The kitchen and baths continue the same style; every detail pursues delicacy and consideration; all elements a house should have are here. This home wakes up the "Peach Garden" in everyone's inner heart; from the perspective of life aesthetics, it skillfully integrates restrained and implicit Chinese culture with exquisite home decorations from all over the world and deduces their essences incisively and vividly.

设计融合了纯正的中式园林院落居住文化，以清雅含蓄的黑白灰作为空间设计的底色，搭配棕褐色的胡桃饰面，稳重低调、不露锋芒。

风格上一脉相承的厨卫空间，在每一处细节上都立求精致与体贴，一个家应该拥有的所有元素，都在这里。这个家唤醒了每个人心底的"桃源"，从生活美学的角度出发，把内敛含蓄的中式文化，与来自世界的精品家居巧妙融合，取二者的精髓，并将其发挥得淋漓尽致。

印迹 IMPRINT

龚志强
ZHIQIANG GONG

景以致远 Far-Reaching Scenery

项目名称 | 宁德阳光园住宅
设计公司 | 福建国广一叶建筑装饰设计工程有限公司
参与设计 | 蔡秋娇、吴鹏飞
项目地点 | 福建宁德
项目面积 | 120 m²
主要材料 | 古堡灰大理石、爵士白大理石、木格栅、木饰面、羊毛地毯等

DESIGN CONCEPT | 设计理念

In this project, the designers intend to apply hand-painted enjoyable painting into designs of every space and combine life habits of modern Chinese to create a new home experience which fits modern living situation. It presents implicit and gorgeous design quintessence of traditional Chinese style and creates a modern, concise and elegant creative space. This kind of design can make people reach a spiritual and clean aesthetic realm so as to burst out more possible associations.

  此次设计中，设计师有意将手绘写意画运用到各个空间设计中，以及结合现代中国人的生活习性，创造出符合现代生活情境之内的家居新体验。在体现传统中式含蓄秀美的设计精髓之外，打造出一个现代、简约和秀逸的创意空间。这样的设计，使人达到一种灵与净的唯美境界，进而迸发出更多的可能性联想。

Into the living room, the designers break conventional design method. The TV wall uses a landscape painting as the main background. Transparent materials on two sides extend to the ceiling through wood line grilling and to the sofa background. Two sides of sofa background use creative method on the basis of the traditional Chinese white space and foil Zen flavors of the space by clean and neat wood lines.

As for the dining room and kitchen, the designers break the traditional closed Chinese kitchen concept and add visual and practical spaces between the kitchen and dining room through the introduction of open kitchen design method. The designers link kitchen with dining room by small bar. Wood veneer ceiling continues the entire wood grilling from the top to walls. There are small size of hanging shelves above the long table to present the owner's pursuit of life quality and sense of design.

来到客厅，设计师打破常规设计手法，电视背景墙上以一幅山水画作为主背景，两侧设计透光材料透过木线条格栅延伸到吊顶，一直延续到沙发背景墙。沙发背景两边则在传统中式留白的基础上做了创新手法处理，以干净简练的木线条手法映衬出空间的禅意风。

餐厅和厨房部分，设计师打破了传统中国的封闭式厨房概念，通过引入西式的敞开式厨房设计方式，增加厨房与餐厅之间的视觉和实用空间。设计师将厨房、餐厅以小吧台形式串联起来，吊顶木饰面则延续整体木格栅形式感从顶延续到墙身，在案几上方做了体量细小的吊架处理，表达居室主人对于生活品质和设计感的追求。

The master bedroom reduces some calmness in ancient style and increases some brightness in Chinese style. The enjoyable landscape bedside background, indistinct lamps and classical and elegant beddings add flavors for the space. The designers ingeniously take use of lighting of the bathroom, combine glass with wood lines to continue ceiling lines of the master bedroom in every independent space and break the certain form that the toilet in a traditional closed bathroom must be fixed, bringing residents joy of life.

The overall designs fully present landscape charm. The mountain and water are endless and continuous. The space uses wood color, white, gray and black as the main tones, collocating with noble blue, which is quaint and elegant. In addition with the poetic natural scenery, there is a fresh and carefree atmosphere in the air.

走进主卧室，少了几分古式的沉稳，多了几分中式的明朗。不论是写意山水的床头背景，还是若隐若现的台灯，还是古典高雅的床品选择，都不自觉地为空间增添了几分的韵味。设计师更是独具匠心，利用卫生间的采光，采用玻璃结合木线条的做法，使主卧室吊顶的线条感延续至每个独立空间，打破传统封闭式卫生间洗手间马桶一定要固定在某种形式上，这样微小的变化给使用者带来生活的乐趣。

纵观整体设计，尽显山水魅力，一山一水，连绵不绝。空间以原木色、白色、灰色、黑色等为主色调，配合高贵的蓝色调，古风古韵十足，又不失典雅。结合诗意般的自然风景画，空气中弥漫出清新、畅快的气氛。

印迹
IMPRINT

杜文彪
BILL DU

小隐隐于野　大隐隐于市

Retreating into the Noisiest Fair is the Greatest Hermit

项目名称｜观承别墅北户
设计公司｜广州杜文彪装饰设计有限公司
项目地点｜北京
项目面积｜662 m²
摄 影 师｜Bill
主要材料｜天然石材、金属、丝布硬包、实木复合地板等

DESIGN CONCEPT ｜ 设计理念

This project is located in central villa area in Shunyi, Beijing, giving the disturbing soul a poetic habitation. Human is the child of nature. Though the outside world is prosperous, it cannot stop the heart to advocate nature and be close to it. Entering the courtyard, the landscape comes to your eyes, which is primitive, simple and natural without a hint of urban vanity.

　　本案坐落在北京顺义中央别墅区，让奔波不安的灵魂，得到诗意的栖居。人是自然之子，纵使外面的世界再繁华，都阻止不了一颗崇尚自然、亲近自然的心。步入庭院，依山傍水之境映入眼帘，古朴自然，不带进一丝都市的虚华。

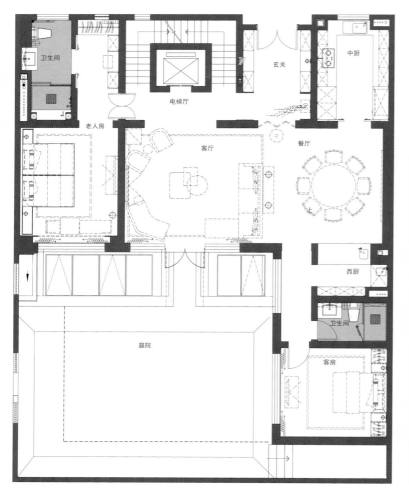

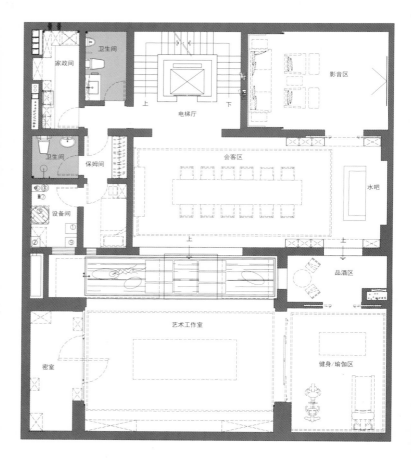

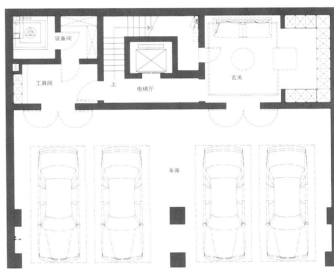

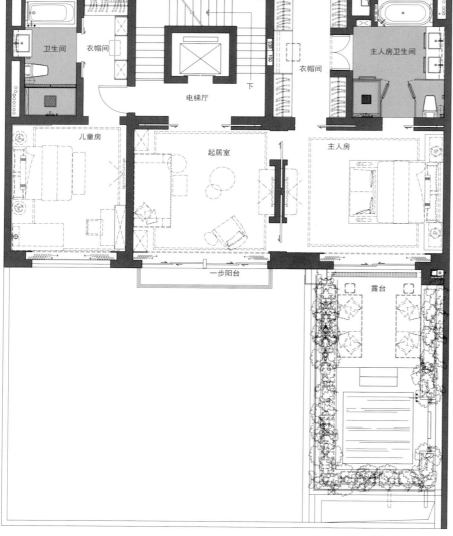

Guancheng Villa is a new exquisite courtyard project, located in the area of "Imperial Gift" converged by "Fangshi Canal" and Wenyu River. From the royal okimichi at the beginning of Tang Dynasty and the Korea's foothold when presented to Chinese Emperor to "Fang Canal" which Emperor Qianlong in Tang Dynasty gave name to commemorate, rich cultural relics make a unique Guancheng Villa in central villa area. Inspired by Beijing Courtyard, the duplexes villa is divided into north and south house types. The north Guancheng Villa is a private courtyard with two layers and integrates exquisite scales of Chinese and Western life philosophy; the design and humanity come down in one continuous line. It takes natural advantages of waters and tree features near Wenyu Riverside and perfectly creates a living state which unities man and nature. The extension of natural elements and the integration of Chinese and Western cultures wake up humanistic feelings both in the past and in future and start the era encounter of man and comfortable residence.

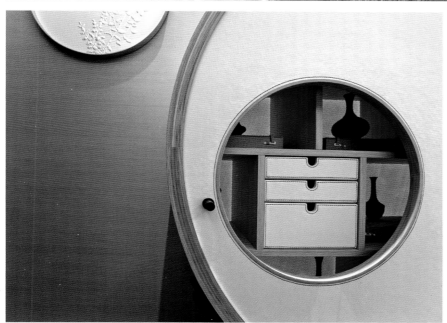

观承别墅，全新精工合院作品，地处"方世渠"与温榆河汇聚而成的"皇脉天赋"板块。从唐朝开始的皇家御道和朝鲜国觐见中国皇帝时的落脚点、到清朝乾隆皇帝御赐姓氏纪念河道"方氏渠"，浓厚的文化遗迹成就了中央别墅区唯一的观承别墅。以京城四合院为灵感，双拼别墅，分为南北两个户型。观承北户，私密庭院，双层空间，融合中西生活哲学的精工尺度，设计与人文一脉相承。借温榆河畔的水系、林木地脉特色等自然优势，完美进阶天人合一的居住境界。自然元素的延伸，中西文化的交融，唤醒过去与未来的人文情怀，开启人与惬意人居的时代邂逅。

This project is guided by Western techniques and Chinese spirits. Modern people are engaged in the busy and noisy city; the tranquility and comprehension brought by Oriental Zen make them immersed in small reclusion that "deep woods that no one knows, where a bright moon comes to shine on me". The designer combines Chinese feelings with Western constructing logic to originally create a Chinese and Western official mansion, manifesting an amiable "home" space temperament. The designer combines open courtyard, uses plants and light and shadow to return to space essence both inside and outside and creates a pure and romantic beauty. Through the cognition of traditional culture, the collision between inheritance and innovative elements derives many beautiful things full of traditional tastes.

　　本案以西技中魂为设计导向，现代都市人在喧闹的城市中忙碌奔波，东方禅意带来静与体悟，让人沉浸于"深林人不知，明月来相照"的小隐生活。将中式情怀与西式建造逻辑相融合，独创出中西合璧的官宅，彰显极具亲和力的"家"的空间气质。结合开放式庭院，利用植物、光影等，内外合一，回到空间的本质，创造出一个返璞归真的美。通过对传统文化的认知，将传承与创新元素相碰撞，衍生出富有传统韵味的美好事物。

Designs of this project take lessons from concepts of ancient landscape paintings. "Bold white space and careful details" is the thought essence throughout the entire project; the essence of white is "simplicity". In the frameworks of Chinese painting, the scenes tend to regard light as fashion, simplicity as elegant and light plain as wonderland. In the tranquil and empty rhythm, presenting the inner rhythm and root essence of man and nature is the suddenly enlightened state between objects and things. White space is a kind of complex and a perfect deduce of pure spirit in contemporary design.

此案的设计借鉴古代山水画的理念，"大胆留白，小心收拾"是本案贯穿始终思想实质，白的本质是"单纯"。在中国画的构架中，画面往往以淡为尚，以简为雅，以淡微为妙境。在恬淡虚无的笔墨韵律中，展示自然与人生的内在节奏与本根样相，即物与事之间的豁然开悟之境。留白，是一种情结，更是当代设计中纯粹精神的完美演绎。

印迹 IMPRINT

OTBHOME

青墙瓦黛 Green Wall and Dark Tile

项目名称｜绿城紫薇公馆·合院
软装设计｜OTBHome
项目地点｜江苏徐州
摄影师｜ingallery™
主要材料｜大理石、木质、布艺、皮革等

DESIGN CONCEPT ｜ 设计理念

Crape myrtle is the city flower of Xuzhou, Jiangsu. Located near Yunlong Lake which is symbolic scenery in Xuzhou, this project faces with mountains and water and is a modern land of idyllic beauty. So Greentown names this project as "Crape Myrtle Mansion".

This project combines city history, local conditions and customs and architectural culture in Xuzhou, integrates them into green mountain and water in Yunlong Lake scenic area and restores Chinese people's final dream to house property.

　　紫薇，江苏省徐州市市花。本案地处徐州标志性风景区云龙湖旁，背山面水，可谓是现代版的世外桃源，绿城将此地产项目定名"紫薇公馆"。

　　本案结合了徐州城市历史、风土人情、建筑文化，并最终融入云龙湖景区的青山绿水之中，还原了中国人对于置业的终极梦想。

Opening the door and coming in, you can see a single house with an independent courtyard. The designer uses buildings, corridors and gray space to enclose the courtyard so as to bring in rockery, waterscape and plants. The overall tone is dark, which is restrained and sedate. Carving partition and enjoyable ink painting match with old tree, deadwood, bottles with plum blossom, which is primitive and tranquil. Designs of the show flat combines with architectural façade with a long artistic conception; every step forward, there is a different scene. White wall and dark tile, carved beams and painted rafters, deep courtyard are like enjoyable landscape painting. In your own house, you can taste fun of natural landscape and gain spiritual joviality. Ornamental perforated window, moon gate and porch connect partitioned spaces so that sights can penetrate into another space, which presents layering changes of the space and achieves separated yet connective effect.

推门而入，独门独院，一户一世界。通过建筑、连廊与灰空间将庭院围合，将假山、水景、植物引入其中。整体色调以深色为主，内敛沉稳。雕花隔断、写意水墨画，再配上老树枯枝、胖瘦梅瓶，古朴宁静的气息挥之不去。其样板的设计结合了建筑外观，意境悠长，一步一景、步移景异。粉墙黛瓦、雕梁画栋、庭院深深，正如写意的山水画，在自家的宅院中，就可以寻求自然山水之趣，获得精神上的愉悦。通过漏窗、月洞门、门廊，使被分隔的空间发生渗透，视线能够穿透到另一空间，显现出空间的层次变化，达到分而不隔的效果。

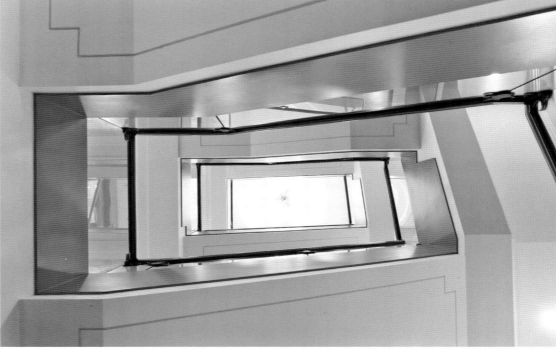

印迹
IMPRINT

王红微
VIVI

半窗疏影恋黄昏
Half Window Flickering the Fading Shadow in the Twilight

项目名称 | 珠海弘泰豪庭
设计公司 | 同心同盟装饰设计有限公司
项目地点 | 广东珠海
项目面积 | 200 m²
摄 影 师 | 蔓视觉影像kader
主要材料 | 铁刀木、黑檀、拉丝古铜、夹丝玻璃等

DESIGN CONCEPT | 设计理念

Since people are tired of minimalism and complicated European style, the retro style of home decoration is gradually enjoyed by people; but it is inevitable that classical Chinese style may be toneless and dull sometimes; so using modern method to reinterpret classical art becomes a new Chinese style.

The creation of neo-Chinese style reflects Chinese people's love to classical Chinese culture and shows modern people's pursuit to elegant, implicit, dignified and refined Oriental spiritual world. It has neither expensive cost nor complicated decoration, keeps classical elements of Chinese style and fits modern people's lifestyle.

　　随着人们对极简主义与繁杂欧式风格的厌倦，家居装饰的复古风正逐渐被人们所喜爱，但是古典的中式风格难免产生单调乏味的情况，运用现代的手法将古典艺术重新诠释，使得成为了一个新中式风格。

　　新中式风格的诞生体现了中国民众对中国古典文化的喜爱，表现了现今的人们对清雅含蓄、端庄丰华的东方式精神世界的追求。它没有昂贵的造价也没有复杂的装饰，既保留了中式风格的古典元素，又符合现代人的家居生活方式。

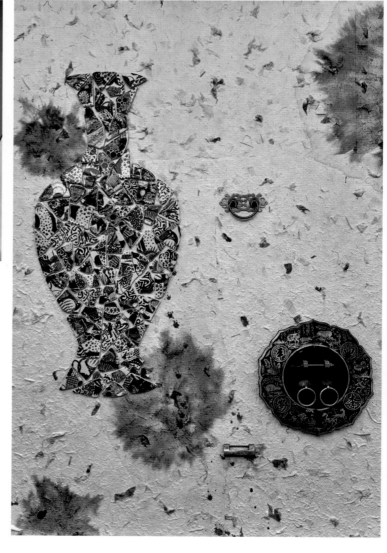

Positioned in neo-Chinese style, this project uses modern method to present traditional culture and element symbols with traditional morals in furnishings, manifesting a kind of precipitation and "relief". The designer carefully masters the relationships between tranquil and active, simple and luxurious, contemporary and traditional, making a perfect combination of classical and modern and a coexistence of tradition and fashion.

　　本案采用新中式风格,将具有传统文化及传统寓意的元素符号运用现代的手法在配饰上呈现出来,表现一种沉淀与"释怀"。设计师在空间中细致拿捏着静谧与活跃、朴素与奢华、当代与传统的关系,让古典与现代完美结合,传统与时尚并存。

印迹 IMPRINT

李益中
YIZHONG LI

藏形露意 Hiding Form and Exposing Meaning

项目名称 ｜ 东莞鼎峰源著别墅
设计公司 ｜ 李益中空间设计
硬装设计 ｜ 范宜华、关观泉
陈设设计 ｜ 熊灿、欧雪婷、欧阳丽珍
施工图设计 ｜ 叶增辉、张浩、王群波、高兴武、胡鹏
项目地点 ｜ 广东东莞
项目面积 ｜ 500 m²
主要材料 ｜ 欧亚木纹大理石、木地板、蓝色妖姬大理石、皮革、木饰面、墙纸、硬包、夹丝玻璃、清水玉等

DESIGN CONCEPT ｜ 设计理念

Dongguan is a prefecture-level city in central Guangdong province, China. Dongguan Origin Villa is located in Dongguan City near Dongguan Botanical Garden with great natural landscape resources.

The project built as the construction standard of South China luxury villa with top landscape sets a new benchmark of Dongguan central city mansion.

　　鼎峰源著别墅位于东莞市南城区五环路边（迎宾公园对面），依临东莞植物园，有独特的自然地理环境，独拥东莞核心城区绝无仅有的"欧洲版"绿色山水资源。

　　按照南中国顶级山水豪宅标准建造，该项目刷新了新东莞的城央山水豪宅标杆。

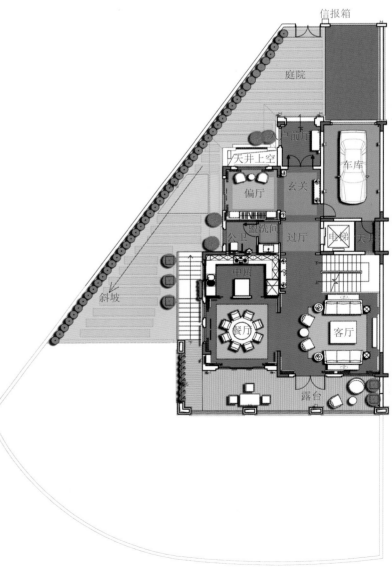

With maximum utilization of resources and user-friendly design, we successfully create a high-quality villa with a taste of life.

We think that the structure is like bones of human-beings, which needs to be tailored well. Different people have their clothes in different styles. Then there should be a different space in relatively suitable style – a landscape villa with oriental charm.

设计师围绕"资源利用最大化,人性化设计,核心空间,项目建筑与周边景观,室内外过渡空间利用"这几大方面来分析该户型,打造一个注重品味,彰显高品质的四层豪宅。

在结构设计方面,设计师认为房子的结构就像人的骨架,必须量体裁衣。不同的人有不同的适合自己的穿衣风格;不同的空间也应有与之相对比较适合的风格面貌,因此定位为富有东方韵味的山水豪宅。

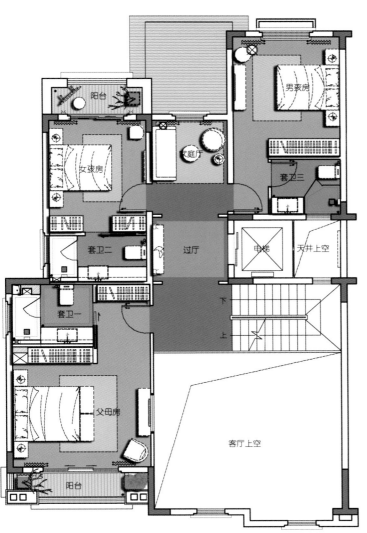
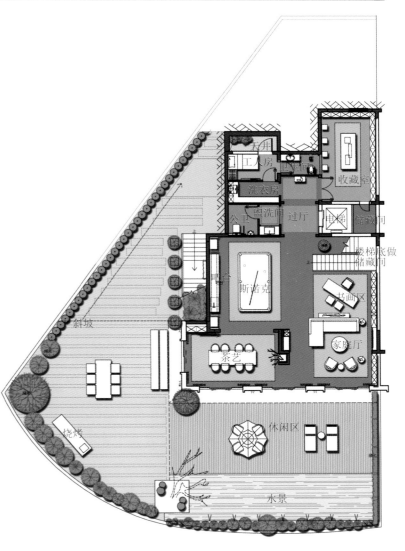

The ground floor has spacious living area perfect for entertaining, which includes painting area, snooker area, a teahouse and a wine cellar.

设计师在前期设计时考虑到户型方正，空间利用率高。负一层是个相对独立而轻松的空间，因此，设计师将这栋豪宅的负一层设计为家庭厅、书画区、酒水吧、斯诺克、茶艺、收藏室、公卫、工人房、洗衣房和储藏间。

The first floor boasts neutral tones throughout and features another generously sized living area which opens onto a relaxing balcony overlooking the tree-lined garden. The second floor contains a stunning bedroom for elderly and two lovely kids rooms. Each space is being fully used.

The generous master room with a relaxing and private terrace enlarges the landscape area on the third floor. The client can come out to the terrace for an air at anytime.

第一层为门廊、玄关、车库，偏厅、公卫、客厅、餐厅、厨房、过厅、露台和天井；第二层为父母套房、男孩套房、女孩套房、小家庭厅和阳台。

第三层则为主人套房、休闲露台、书房和过厅。设计师将露台纳入主卧使用，扩大了主卧的景观面积，同时增添了生活的趣味性。

Modern, simple and rich are our design philosophy. We use simple colors and pursuit the form of concise, while focusing on comforts and emphasizing sense of design. We explore the innovation of oriental elements to create a modern urban space with oriental culture.

  设计师以现代设计手法，简洁而丰富的理念为基础，应运干练利落的色调并追求形式简练的统一，同时注重舒适性，强调设计感。探索对东方元素的吸取与创新，营造一个具有东方文化气息和现代都市并存的空间。

印迹
IMPRINT

戴勇
ERIC TAI

素雅空间
Simple and Elegant Space

项目名称 ｜ 名门紫园29号楼GC2东户型样板房
设计公司 ｜ 戴勇室内设计师事务所
项目地点 ｜ 河南郑州
项目面积 ｜ 336 m²
摄 影 师 ｜ 陈维忠
主要材料 ｜ 云灰云石、柏斯高灰云石、金镶玉云石、雅士白云石、直纹白玉云石、英伦玉云石、银影木饰面、灰鳄木饰面、缅甸胡桃木地板、加拿大枫木实木地板、墙纸等

DESIGN CONCEPT ｜ 设计理念

This project is a suit of big flat mansion show flat Ancestry Purple Garden from Ancestry Reality located in Zhengzhou Zhengdong Green Expo Garden. It pursues that "only in Purple Garden can you review all the Zhengdong scenery" and tries to create high quality city standard.

本案是位于郑州郑东绿博园区域的名门地产项目名门紫园的一套大平层豪宅样板房。有着"阅尽郑东唯紫园"诉求的名门紫园项目，力图打造高品质的城市标杆。

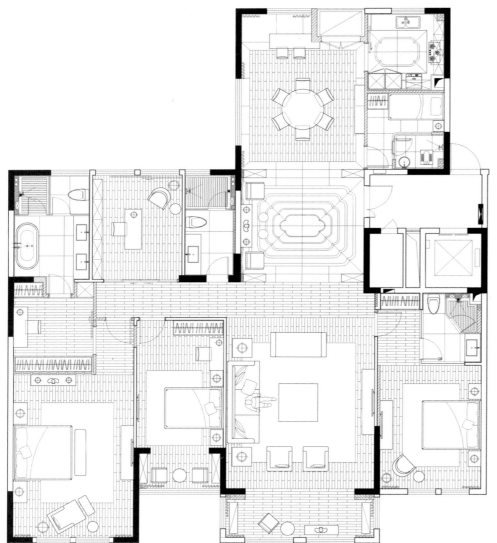

The designer properly divides the entrance area into hallway, maid's room and cellar, which ensures moderate area of the hallway, adds more functions and creates a psychological feeling of bating followed by developing caused by proper compression of the hallway.

设计师把入口处的区域合理划分成门厅、工人房、藏酒室三个部分，在保证了门厅面积适度的同时，增加了更多的使用功能，并通过适当压缩门厅面积形成先抑后扬的心理感受。

The hallway separates dining room from living room and makes them relatively formal and independent. The dining room has a bar and a capacious Chinese and Western kitchen. The living area brings original balcony into interior to form formal living room and golf practice area. The living room has easy access to children's suite, elder's suite, guest room and master suite which is equipped with independent working area and spacious cloakroom with nearly 40 square meters. Every functional layout is fluent and each individual space is square and spacious.

It chooses beige, warm gray and gold color combinations to create an exalted and elegant color atmosphere. Floors of public areas are covered with Turkish cloud gray marbles. Hard decorations adopt gray shadow wood, leather hard coverage, wired craft glass, Burmese walnut and Canadian maple inserted stripe floor. Walls and floors of the bathroom are paved with Italian Perth high gray marbles with a modern wash basin. The wash basin and mirror use stainless steel to close up.

门厅分隔出餐厅与会客厅两大主要区域，让两个空间都相对正式和独立。餐厅区域设有吧台、宽大的中西结合式厨房。客厅区域把原有阳台纳入室内，形成正式的会客厅和高尔夫练习区。从会客厅可以到达小孩房套间、长辈房套间、客房及主人套间。面积接近40平方米的主人房里设有单独的办公区和宽敞的衣帽间。各功能布局流畅，各个单独空间方正宽敞。

色彩选用了米黄、暖灰与金色的组合，营造尊贵典雅的色彩氛围。公共区域地面用土耳其云灰云石大面积铺贴，硬装选用灰影木、皮革硬包、夹丝布工艺玻璃、缅甸胡桃木与加拿大枫木镶嵌条纹地板等物料。洗手间墙地面选用了意大利柏斯高灰云石，现代的洗手台设计，洗手台及镜面用不锈钢做细节的收口。

The interior design and art furnishings of this show flat are designed by Eric Tai Design Co., Ltd. Many pieces of furniture and lamps are customized by the designer to ensure the novelty and uniqueness of soft furnishings. Specially made "cloud" and "peony" titanium stainless steel art hangings in the living room and master bedroom and American brand global views branch chandelier and custom-made forest carpet in the dining room are hints of location of the project. Mr. Eric Tai personally and carefully chooses every piece of display product and adjusts display details on site to ensure perfect execution of designs.

样板房的室内设计及艺术陈设由戴勇设计师事务所全程负责，多件家具、灯具均为设计师原创设计并定制，确保软装陈设的新颖和独特。客厅及主人卧室的特别设计制作的"云"及"牡丹"钛金不锈钢艺术挂饰，餐厅选配的美国品牌global views的树枝吊灯和定做的森林地毯都是对项目所在地理位置的暗示。戴勇先生更是亲自悉心选择每一件陈设产品，并现场调整陈设细节，确保了设计的完美执行。

印迹 IMPRINT

高文安
KENNETH KO

山水之间 诗意栖居
Poetic Residence Between Mountains and Waters

项目名称 | 桂林漓江郡府别墅
设计公司 | 深圳高文安设计有限公司
项目地点 | 广西桂林
项目面积 | 257 m²
摄 影 师 | KKD推广部
主要材料 | 意大利冰玉、大理石、实木、玻璃、布艺等

DESIGN CONCEPT | 设计理念

Oriental soul lies in landscape and the poetry hidden in the waterside of Lijiang River is waiting quietly for its discoverers.

Landscape is spiritual aesthetics unique to Chinese. The Lijiang River has no equal in this world. Spiritual mountain and beautiful water breed unique human philosophy of Lijiang residence. It is not for insisting a style or trend in a certain period but for pursuing beauty and tranquility, not for pursuing formalistic romance but for inner profusion and life attitude keeping up with the times.

　　东方的灵魂在山水，隐于漓江水畔的诗意，静待它的发现者。

　　山水是独属于中国人的精神美学。漓江，一带山河，举世无双，灵山秀水孕育了漓江郡府独一无二的人居哲思，不为坚持某一时期的风格或潮流，而是为了追寻美和宁静，不追求形式主义的浪漫，而是内心的丰盈跟与时俱进的生活态度。

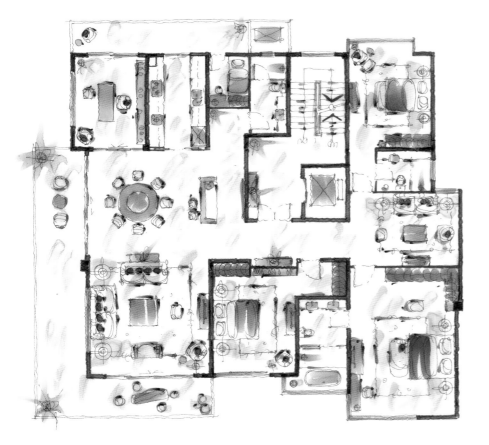

FULL AND DELIGHTFUL CHINESE CHARM
| 醉墨淋漓中国味

Mr. Liang Qichao said: "human must live with fun so that life has its value."

Entering the villa, whitebait furnishings on the wall, ink painting porcelain and hanging pictures of the bar and the painting *Pine Trees and Waterfall* from modern painter Li Huayi with typical Oriental serenity and elegant aesthetic characteristics bring profound and pleasant tone to interior, which renders feelings of enjoyable landscapes.

　　梁启超说："凡人必常常生活于趣味之中，生活才有价值。"

　　步入别墅，墙面的银鱼装饰，酒吧的水墨花器、挂画，玄关现代画家李华弋具有典型东方静穆、典雅美学特征的《小景松瀑》图，意境深远的写意格调翩然入室，渲染纵情山水的情趣。

WHEN THE MOUNTAIN FLOWERS ARE IN FULL BLOOM ｜ 待到山花烂漫时

What is beauty? For the designs of Lijiang residence, it is that I can have everything and have nothing. Between mountains and waters, all makes complicated things simple. Naturally it is beauty.

The interior designs of living room and dining room adopt natural colors without breaking the aura brought by Guilin elegant landscape. The designer boldly replaces most of walls by French windows which have easy access to scenery. Thus natural scenery becomes the largest leading role. Outside the window, there are green mountains and rivers. Inside the window, there are "mountain flowers in full bloom, twitters of birds and fragrances of flowers".

何谓美？对于漓江郡府的设计而言，是我可以什么都有，也可以什么都不要，山水之间，一切化繁就简，自然而然，就是美。

客餐厅，室内设计尽量采用自然的色彩，不去破坏桂林秀美山水带来的灵气。极具魄力地拆除了大部分墙体，取而代之的是适宜观景的落地窗，自然风光成为最大的主角，窗外青山绿水，窗内"山花烂漫，鸟语花香"。

A SENSE OF EXQUISITENESS BECOMES HEAVY IN TRANQUIL SPACE | 一片幽情静处浓

The designs of side hall follow concise proposition of modernism. Lotus in the colorful vat on the corner presents tranquil Zen mood like Chinese landscape. The creative painting on the wall made of wood shavings with spray colors makes the space know more about human changes in temperature.

Wood grilling of the study conveys Oriental life feelings. Wood furniture and floors in the same space present Chinese solemnity, which is comfortable and natural.

偏厅，设计遵循现代主义的简洁主张，墙角彩缸里的莲蓬又带着几分中式枯山水的清寂禅意，墙上以木头刨花喷色制成的创意挂画，让空间更懂人情冷暖。

书房，木格栅传达东方对生活情调的雕琢，木质家具、地板在同一空间中渗透中式肃穆，舒适而不做作。

FROM NOW ON, WHEN TIME AND MOONLIGHT ALLOW | 从今若许闲乘月

Chinese style is the precipitation of time, the extension of history and the combination of ancient feelings and modern sentiments.

The master bedroom continues the recluse feelings of "plum as wife and crane as son" pursued by refined scholars since ancient times. The traditional Chinese painting on the background wall makes up humanistic connotations which modern space is lack of. It seems that living in seclusion in the mountain forest, far away from the hustle and bustle and planting plum and breeding crane are indifferent and comfortable.

中式是时间的沉淀，历史的延伸，古老情怀与现代情操的结合。

主卧室延续了自古以来文人雅客追求的"梅妻鹤子"隐逸情怀，背景墙上的国画补足了现代空间欠缺的人文底蕴，仿佛隐居于深山茂林，隔离尘世喧嚣，植梅养鹤，清高自适。

The tone of parents' room is simple, elegant and warm, which is good for the elders to calm down and soothe the nerves. The proudly blooming thin plum on the background spares no effort to bring people close to Chinese poetry. Under the moonlight, there seems to be a secret fragrance.

As for the subaltern room, the designer deeply knows the truth of "the more national, the more worldwide". Life space becomes the stage to show culture. The hanging painting made of Miao silver and the classical floral droplights gently relate the splendid culture of Chinese ethnic minorities.

Guilin Lijiang residence is a landscape residence built on a spiritual level. Living in seclusion near Lijiang riverside, picturesque mountains seal up many poetic qualities and flavors. The footpath strewn with fallen blooms is not swept clean; my wicket gate is opened but for you today.

父母房，空间色调素雅而温暖，更有助于老年人静气安神。背景上傲然绽放的瘦梅，则不遗余力地带人走近诗意中国，月夜下似有暗香浮动。

次卧，设计师深谙"越是民族的，越是世界的"这一道理，生活空间成为展示文化的舞台，苗族银饰做成的挂画，古典花灯风情的吊灯，将中华少数民族的灿烂文化娓娓道来。

桂林漓江郡府，是建立在精神层面上的山水居所，隐于漓江水畔，如画的山色中封存多少诗意，花径不曾缘客扫，蓬门今始为君开。

印迹 IMPRINT

张炜伦
KAREN CHANG

雅藏 Elegant Collection

项目名称 ｜ 悦湾别墅
设计公司 ｜ 卡纳设计·上海
参与设计 ｜ 褚则则、付莉莉、梁宇华
项目地点 ｜ 重庆
项目面积 ｜ 350 m²
摄 影 师 ｜ Sam
主要材料 ｜ 木材、大理石、玻璃、金属等

DESIGN CONCEPT ｜ 设计理念

This project takes advantage of storey height to make stereo space cutting as the design focus. One floor upward, one can overlook the hanging down chandelier, as if the light and shadow of moon entering into the water, which is poetic. Looking straightforward, the upstairs and downstairs stack and scatter, indicating the geography of mountain city Chongqing.

Dark marble and wood materials constitute sedate main tone, conforming to aesthetic taste and humanistic spirit of the owner. The ceiling with concise lines and carpet and pictures with random ink painting set off each other like the sky and water and couple hardness with softness. The processing of stair handrail presents contemporary carve spirits. The up and down lines are like the folding corridors in classical garden, which makes a piece of land vivid abruptly and full of chic taste.

　　本案借助层高上的优势，将立体空间的切割作为设计重点。再上一层，俯瞰吊灯错落垂下，似明月光影入水，诗意盎然。而平视之时，上下楼之间的层叠错落，暗合山城重庆的地理环境。

　　深色大理石和木材构成沉稳的主色调，应和主人的审美趣味和人文精神。线条简约的吊顶和水墨笔意纵横的地毯、挂画，如天水相映、刚柔并济。楼梯扶手的处理是当代雕琢精神之所在。线条高低起伏，犹如古典园林回廊折叠，令一方天地骤然生动起来，气韵别致。

Following traditional essences of Chinese style, the cutting of every functional space is very square. The furniture is mainly in classical patterns, such as round stools, concise floor lamps and magnificent marble table, manifesting the temperament of the literary person. In addition, the designers specifically use soft lamps and romantic flowers in details, adding some feelings of spring and making tranquil and comfortable life atmosphere flow quietly.

沿用中式传统精髓，各功能空间之间的切割极为方正。家具的选择多以古典式样为模板，诸如圆凳、简约立灯和铿锵有致的大理石桌，彰显名士气度。此外，设计师特意在细节处用柔和灯具和浪漫花艺，增添几许春意，让安宁舒适的生活氛围，悄然流转。

印迹 IMPRINT

谢雨时
ULYSSES

源于内心 写意山水
Enjoyable Landscape Inspired by Inner Heart

项目名称｜玥珑湖生态城A1栋联排别墅
设计公司｜广州欧申纳斯软装饰设计有限公司
项目地点｜广东湛江
项目面积｜255 m²
摄 影 师｜谢雨时

DESIGN CONCEPT ｜ 设计理念

We often describe Chinese culture as extensive and profound and are sincerely proud of it. The five-thousand-year bright civilization seems to stay forever in the past, being in China and far away from China. The designer wants to make some changes by combining Chinese context with local culture in Zhanjiang and using contemporary design technique to sublimate the tone of Chinese style. Landscape is good for one's health; using landscape to please life is the way to keep in good health since ancient times. Modern neo-Chinese style with Chinese landscape elements creates an elegant, dignified and comfortable art space which is full of artistic connotations by sedate mountain and leisure cloud.

我们经常用博大精深来形容中国的文化，同时由衷地为中国文化感到自豪和骄傲。沿袭了5000年璀璨文明好像永远停留在过去，身处中国却远离中国。作为一个设计师希望能做出一点改变，用中式语境结合湛江当地文化，以当代设计手法让新中式的格调再次升华。山水养生，用山水愉悦生命，这是自古以来的养生方式。现代新中式的设计风格，融入中国山水元素，以山之沉稳，云之悠闲，营造优雅、厚重、舒适，富有艺术底蕴的艺术空间。

Design comes from life and also one's state of mind. The designer wants to convey a living mood which has nothing to do with joy and sorrow, wind and moon, but a kind of easy return with blue mountains and poetic paintings.

设计源于生活,也源于人的心境。设计师要传达的是一种生活的心境,如此心境,无关悲喜,无关风月,只是一种蓝山画意式的从容回归。

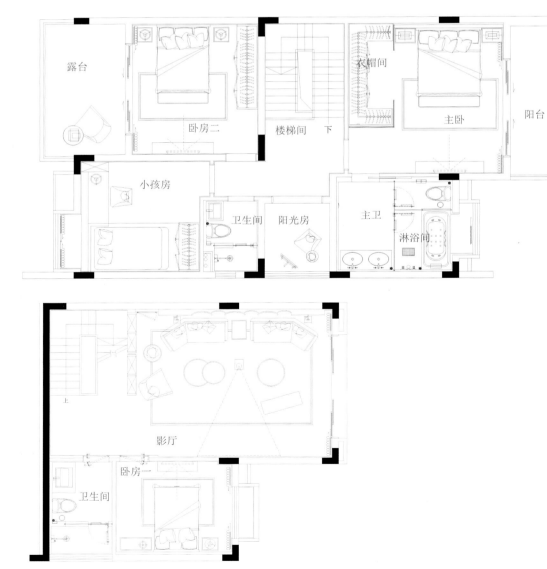

IMPRINT 印迹

张凡
张磊
FAN ZHANG
LEI ZHANG

水墨山行 Ink Painting Mountain Trip

项目名称 ｜ 福州三盛地产滨江国际
空间设计 ｜ 深圳大森设计有限公司
软装设计 ｜ 深圳叶迹艺术顾问有限公司
项目地点 ｜ 福建福州
项目面积 ｜ 114 m²
摄 影 师 ｜ 大斌室内摄影
主要材料 ｜ 伯爵米黄大理石、金镶玉大理石、爵士白大理石、黑钛金、香槟金拉丝不锈钢等

DESIGN CONCEPT ｜ 设计理念

The inspiration of the space comes from *Mountain Trip*. Traditional Oriental concepts pay attention to artistic conceptions which are personal feelings of "traveling in a pictorial world". Designers use varying sceneries with changing view-points and cadence techniques to create extreme artistic conceptions and dignified senses for the space, manifesting restrained humanistic temperament and connotation of the entrance. Large areas use ivory beige as the main tone, with bronze line frames enriching details. Ceiling with concise black frames matches with ink painting and sedate and magnificent Chinese furniture, creating visual effects by modern aesthetic pursuits and abandoning complicity and noises, which is flowing and tranquil, profound and exquisite, and endows the rich and orderly space with quaint artistic conceptions.

　　空间的灵感来自《山行》，传统的东方理念讲究的是意境，是"人在画中游"的亲历感受。设计师运用步移景异、抑扬顿挫的手法来营造空间的极致意境和尊贵感，彰显出府门内敛的人文气质和涵养。大面积采用象牙米色作为基调，用古铜线条收边来丰富细节。简练黑色边框的天花，衬以水墨画与沉稳大气的中式家具，以现代审美诉求进行视觉营造，去除繁杂与喧嚣，流动却又静美，博大却不乏精细，将丰富、有序的空间演绎出古雅的意境。

The world famous architect Tadao Ando said: "The so-called tradition is not visible form but the spirit which supports the form. I think, absorbing this kind of spirit and using it in modern society is the real meaning of traditional inheritance."

世界建筑大师安藤忠雄说:"所谓传统,不是看得见的形体,而是支撑形体的精神。我认为,汲取这种精神并在现代活用,才是继承传统的真意。"

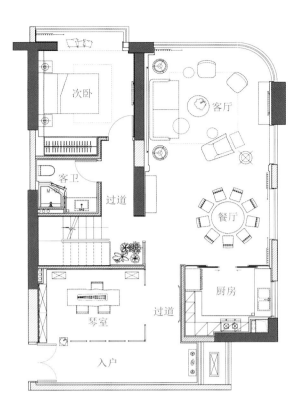

印迹
IMPRINT

李超
CHAO LI

相遇时光
Meeting with Time

项目名称 ｜ 翡翠松山湖滨湖花园C户型样板别墅项目
设计公司 ｜ 深圳市圣易文设计事务所有限公司
项目地点 ｜ 广东东莞
项目面积 ｜ 620 m²
主要材料 ｜ 玛莎蒂灰大理石、孔雀蓝玉大理石、意大利冰玉大理石、年轮大理石、路易斯白大理石、墙纸、木饰面、地毯等

DESIGN CONCEPT ｜ 设计理念

In Songshan Lake, one will have a feeling very different from "steel forest" in big cities. Clean sky, green plants, warm air, everything is very real within easy access. The designer is struck by the beauty and artistic conception of Songshan lake and deeply into it, so every detail design in Binhu Garden Villa is very delicate. Dongguan, on the one hand, is a booming city like the sun, on the other hand, is a city with a long history. At the beginning of the design, we give the space more expectation and hope she can be ahead of the trend and has cultural charm, Western romance and Oriental backbone. The space atmosphere of mixed Chinese and Western style has a strong tension yet is not too flaunty. We hope that everyone who enters into this space can feel the beauty of time collisions and cultural differences.

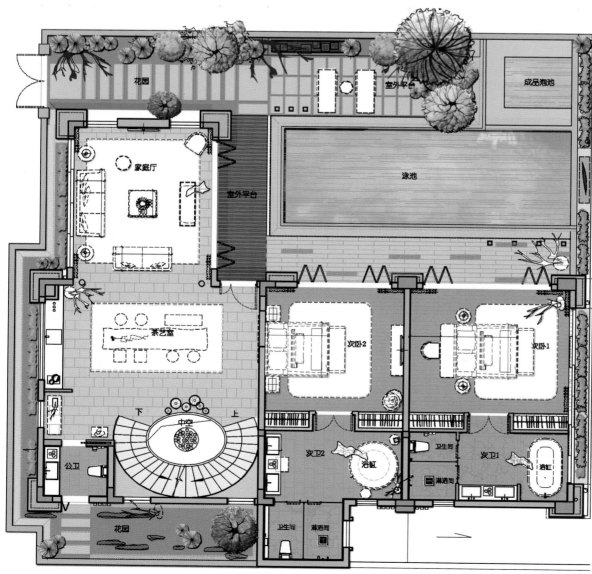

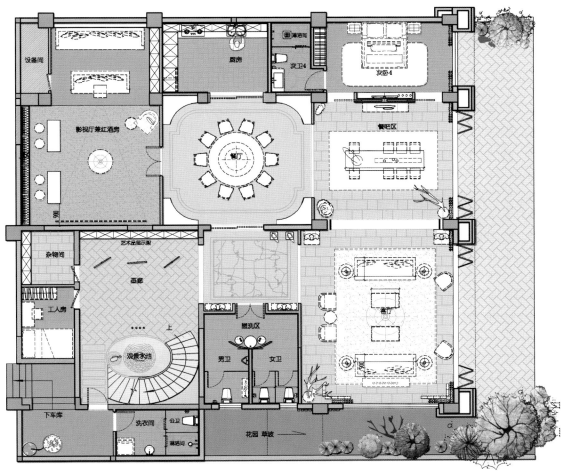

　　身处松山湖，一种与大城市"钢铁森林"截然不同的感受。天空的洁净，植物的新绿，空气的温润……一切都十分真实，触手可及。设计师用于滨湖花园别墅的每一个精致细节，都源自于对松山湖的醉心与感动。像莞城一样，一方面是个朝阳般蓬勃发展的城市，另一方面是一个有着深厚历史故事的城市。设计之初，我们寄予空间厚望，希望她能领先于潮流，亦要有文化的韵味，有西方的浪漫也有东方的骨气。中西混搭的空间氛围，有着强大的张力却不过分的张扬。希望让每一位进入这个空间的人，都能感受到时光的碰撞与文化差异之美。

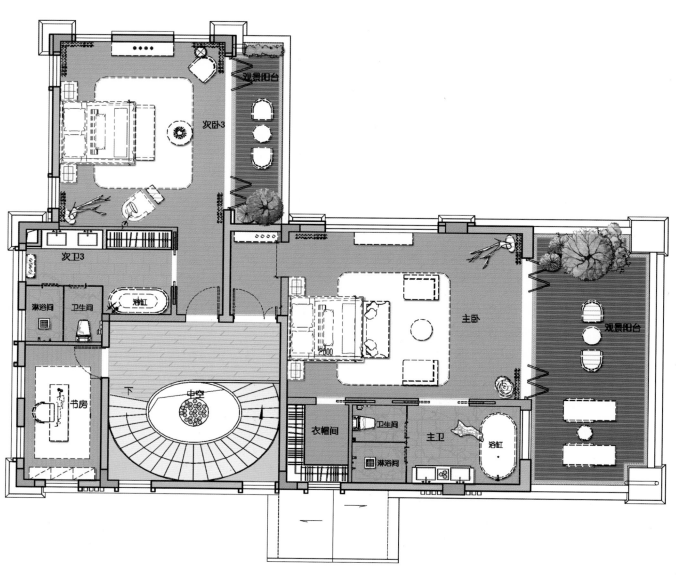

IMPRINT 印迹

岳蒙
MENG YUE

White Space 留·白

项目名称 | 云谷样板间
设计公司 | 成象设计
软装公司 | 成象软装
项目地点 | 山东泰安

DESIGN CONCEPT | 设计理念

Playing walnut, playing calabash and loving seal cutting,
He is a real craftsman in spare time.
Reading Lu Xun's novels, drawing traditional Chinese painting and deeply researching *Diamond Sutra*,
He says life should have a little aura.
Accompanying with books, he keeps romantic,
Understanding Zen, he loves Oriental white space,
Tasting a cup of tea, producing different flavors.
Taking a sip is better than getting drunk.
Once expected waves of fate,
Now finds that,
The most spectacular scenery in life,
Is the inner calmness and leisure.
Read books, recite poems, rub inks and draw pictures,
Watch clouds gathering and spreading and time changing,
Playing the piano, meeting with friends in melodious atmosphere,
If the heart is free and unfettered, one can see sunsets and hills in the house.

盘核桃、玩葫芦、爱篆刻，
闲暇时他是个十足的手艺人。
读鲁迅，画国画，深研《金刚经》，
他说生活里要有一点灵气。
与书为伴，他保留浪漫，
参悟禅语，他爱东方的留白，
品一杯茶，百味尽生。
小酌尤胜一晌贪欢。
曾渴望命运的波澜，
而今才发现，
人生最曼妙的风景，
是内心的淡定与从容。
可读书吟诗研墨挥毫，
可观云起云卷光阴变幻，
操琴会友，琴瑟悠扬，
心若逍遥，于居中亦能看遍夕阳山丘。

Hollow-out, log and ink painting integrate skillfully; soothing Oriental artistic conception is the owner's unique feeling. Being indifferent to show ambitions and being tranquil to achieve goals, dignity and Zen are the temperament and interest of Chinese style. Whispering with the fleeting time and scattering flowers and glasses in the house, we wish to find a sweet smile and put all beautiful things here.

Chinese furniture with nostalgic feelings integrates ancient meaning with the owner's mind. Warm window sides and cozy corners are full of amorous feeling; the artistic conception with flowing water is exactly like the story of the family. The ethereal sandalwood accompanies your reclusive soul and falls on the Oriental Zen state. Tea can help you taste different life flavors; book can help you find spiritual conversion; with a brunch of sunshine, the study is a place to meditate. Simple colors and soft cloth are the owner's pursuit of simplicity, which can make the owner forget all the complicities. Red and beige are full of young fervor and brightness; the single sofa near the bed provides a place to quietly enjoy alone.

镂空、原木、水墨巧妙兼容，舒缓的东方意境是主人特有的情怀。淡泊明志，宁静致远，庄重与禅意是中国风的情致。细语流年浅唱岁月，将花花草草散落于家中，只愿寻一片嫣然，将美好的一切都凝固在这里……

寄托怀旧情绪的中式家具，将古意与主人的心绪融合。温暖的窗边，惬意的角落，处处皆风情，细水流长的意境正如家的故事。悠悠的檀香，伴着你想隐遁的灵魂，归于这一处东方禅境。茶可品尝人生百味，书可以找回心灵皈依，轻拥一米阳光入怀，书房便是冥想天地。简单的色彩与柔软布艺是主人对质朴的追求，把所有的繁复遗忘。红色偎着米色，有种青春的炽热与明媚，床边一套单人沙发，可静静享受独处的精彩。

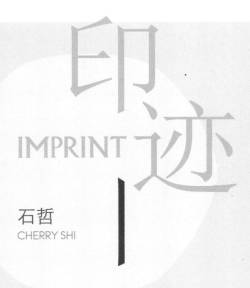

印迹
IMPRINT

石哲
CHERRY SHI

云裳荷影　落红拂柳
Shadows of Cloud and Lotus, Flowers Falling Through the Willow

项目名称｜绿城百合花园会所
陈设设计｜北京中合深美装饰工程设计有限公司
参与设计｜廖一莎、郭小雨
项目地点｜山东济南
项目面积｜400 m²
主要材料｜屏风、装饰画、墙纸、花艺等

DESIGN CONCEPT ｜ 设计理念

To be honest, Chinese design trend develops very rapidly. No matter conception or ideology, people pay more attention to dig the spiritual needs and humanistic care of the design itself. In the design, when facing a group of elites who pursue identity and recessive business needs, the designers take care of their needs, explore and interpret Oriental culture, present a new appearance and bring a more profound spiritual aftertaste. At the beginning of the design, the designers make premises for the design of the villa that refusing the "inane without anything" of the villa configuration and using extreme and pithy designs endow every floor of the 400 square meters villa with special functions and correspondent funny connotation so as to make it to be a custom-made and high-end social club in Jinan as well as a villa for living. It is an exclusive and tranquil space, an extension of family living space, and a dream worshiped by people.

　　不得不承认，中国设计思潮发展非常迅速，无论是观念还是意识形态，人们对设计更加注重挖掘能对应精神认同层面的需求和设计本身应有的人文关怀。在设计中，面对一群寻求身份认同、有隐性商务需求的精英群体，设计师关照其需求，对东方文化进行发掘和演绎，让人耳目一新并带来更深层的精神回味。在设计之初，设计师为该别墅的设计注入了前提：拒绝豪宅配置下的"空洞无物"，用极致而精粹的设计，为这套400平方米的别墅会所每一层都赋予其独特功能和对应的趣味内涵，使之进可为济南城市峰层定制的高端社交会所，退可为别墅居住空间，这是一个专属的静谧空间，是家庭生活空间的延伸，也是人们心中渴望被人尊崇的梦。

There are living room, dining room and kitchen in the first floor. The main tone of this floor is elegant white and beige. Gold screen promotes the temperament of the space. Bright yellow and fresh gray blue intersperse in the space. The collocation and reconstruction of Chinese and Western elements create an artistic exclusive style.

Large area French window in the living room brings outside natural scenery into interior. The furniture is upright and straight contemporary style, while the screens and decorative paintings manifest Chinese style with enjoyable ideal and splash-ink as the theme, which freely interprets Oriental essences. Imported velvet fabrics and crystal ornaments integrate with metal temperament, deducing a highlight of the space.

Designs of the dining room take dining etiquettes of large family and literati feelings into consideration. In the beige tone, bright crystal droplight, splendid round marble table, customized dining chairs, metal screen and contemporary Chinese porcelain and antiques communicate in the same space, indicating unique aesthetics and manners.

一层是客厅、餐厅和厨房。该层空间以素雅的白色和米色为主色调，金色屏风提升空间气质，明亮的黄与清新的灰蓝点缀其间，中西方元素的搭配与重组，构筑出空间艺术氤氲的专属格调。

客厅大尺度的落地玻璃窗将户外绿意盎然的自然风景引入室内，家具样式的选择以刚直挺拔的当代风格为主；而屏风和装饰画却带入中式风情，以写意为理想，泼墨为主题，收放自如地诠释了东方精髓；进口的丝绒布料、水晶饰品与金属的气质融合，演绎成空间的妙笔。

餐厅的设计兼容了大户宴客礼仪和文人情怀，在米色的基调中，璀璨的水晶吊灯、气派的大理石圆桌、特别定制的餐椅、金属屏风及中国当代瓷器、古玩在同一空间对话，隐喻独到审美和气度。

Chinese red and black collocate perfectly in the underground first floor with exquisite modern Chinese style. Customized tassel droplight, super-scale ink painting, dark marble table, soft silks and satins and modern furniture interweave in the space, gaze, echo and resonate with each other, reinterpreting a new posture to the world and depicting a magnificent and luxurious space.

There are two reception areas in the second floor, continuing tone in the first floor yet with unique tastes. It injects sedate decorative details into contemporary style, skillfully integrates classical charms with modern designs, makes the space full of rigorous and pride, reveals the power of sense of ritual and scarcity and conveys contemporary Chinese artistic aesthetics.

地下一层中国红和黑色完美搭配，弥漫着精美绝伦的现代中式风情。特别定制的流苏吊灯，超越尺度的巨幅写意画，深色系大理石桌，丝滑柔软的绸缎，现代家具，这些元素交织在一起相互凝视、呼应、共鸣，向世界重新阐释了华夏新姿，描绘了一个瑰丽奢华的空间。

二层布局了两个宴客空间，延续一层的格调却又各有风情。在当代风格之中，植入稳定气派的装饰细节，将古典韵味与现代设计进行巧妙融合，空间充满严谨和骄傲，透露出仪式感和稀缺感的力量，表达当代中国艺术审美。

As business and life function areas, the third floor is set with tea room, meeting room and bedrooms. Designs of the tea room uses new Zen method to make a collision between Oriental "spirit" and Western "form" and to produce a dialogue. Concise lines match with light elegant wood, displaying a space aesthetics with latitude and depth and endowing the space with endless artistic conception.

The bedroom is elegant and beautiful with beige as the main tone and dark blue interspersing in it. Wallpapers with flowers and birds salute to Chinese style. Unique decorative furnishings and floral arrangements make the whole bedroom elegant and charming.

There are billiards room, dance room and children's playing area. In order to promote temperament of the space, the soft decoration matches cool noble dark blue, bright yellow and elegant white with sedate wood, creating a funny activity space.

三层是完备商务与生活功能区，设置茶室、会议室和卧室。茶室的设计以新禅意的手法将东方的"神"与西方的"形"相碰撞并产生对话，极简的线条和淡雅的木色相搭配，具象地展示了更具纬度和深度的空间美学，让空间拥有无限的意境。

卧室的雅致和美如影随形，米色为主色，深蓝色铺陈其间，花鸟的墙纸向中式致敬，造型独特的装饰艺术摆件和花艺使整个卧室清雅迷人。

四楼是台球室、舞蹈房和儿童玩乐空间，为了提升空间气质，软装陈设以冷艳高贵的深蓝、明亮跳跃的黄和雅致的白应对沉稳的木色，营造了一个意趣盎然的活动空间。

IMPRINT
印迹

巫俊逸
JUNYI WU

清风朗月照人眠
Sleep with Cool Breeze and Bright Moon

项目名称 ｜ 苏州雅戈尔紫玉花园
设计公司 ｜ 午逸宸建筑设计顾问有限公司
项目地点 ｜ 江苏苏州
项目面积 ｜ 450 m²
摄 影 师 ｜ 温蔚汉

DESIGN CONCEPT ｜ 设计理念

Concise and modern hard decoration and soft decoration full of Chinese flavor create a modern and Chinese atmosphere, which is the highlight of this project. Peacock blue and rosefinch red are the center of the decorative tones and bring a piece of gorgeousness for the implicit space. Hand-painted wallpaper presents jasper and elegance of Suzhou; Chinese furniture in bright colors adds vitality for the whole space; a set of droplights endow the space with sense of rhythm; neo-Chinese style which breaks the routine often gives people surprises; the proper treatment of details is the transformation from quantitative change to qualitative change and is the basis to make the entire show flat exquisite. On soft decoration display, the designer ingeniously matches every small article with the color and style of the furniture and space, which extends the artistic conception and taste and makes everyone who stops here to feel its own unique temperament exclusive to this space. Now more and more people need spiritual peace, want to come back from the blind material pursuit, start to pay attention to cultural connotations and pursue natural and comfortable life state. This kind of simplicity, frankness and peace is their real life appearance.

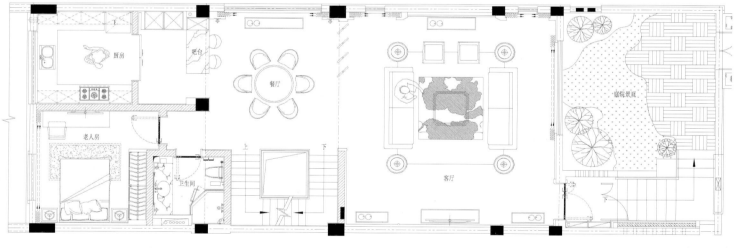

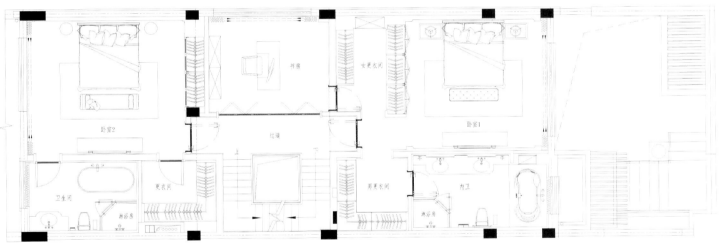

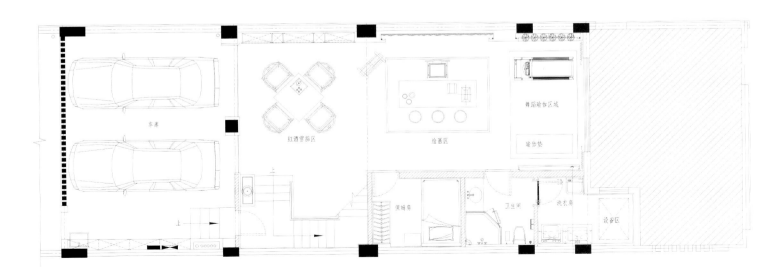

简洁现代的硬装造型，极富中式韵味的软装，打造现代中式气息的空间氛围是本案的亮点。孔雀蓝、朱雀红是空间装饰色调的中心，为空间的含蓄带来了一抹绚丽。墙上的工笔手绘壁纸向人们展示着属于苏州的碧玉秀雅，偶尔混入色彩亮丽的中式家具为整个空间增添生机，组合吊灯的运用赋予空间节奏感，打破常规的新中式总能给人更多惊喜，细节上的稳妥处理是量变到质变的转换，是让整个样板房变得精细的基础，而在软饰的陈设上设计师也是独具匠心，用每一样小物件融合家具与空间的色调与风格，延展了意境与品位，让每一个驻足过的人感受到这个空间专属于自己的独特气质。现在的人们越来越需要心灵的宁静，并希望从盲目的物质追求中退回来，开始注重文化内涵，追求自然、舒适的生活状态，而这种淳朴、率真、平和才是人们真实的生活面貌。

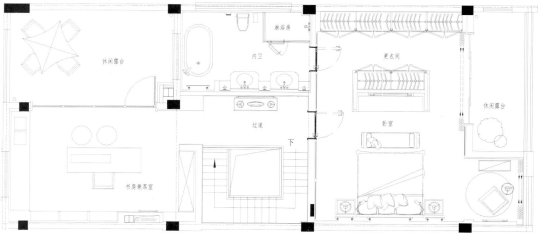

印迹 IMPRINT

上海多姆设计工程有限公司
SHANGHAI DOMUS DESIGN & ENGINEERING CO., LTD

闲来话尽花满庭
Spare Time at Courtyard Full of Flowers

项目名称 | 申亚花满庭S1别墅
项目地点 | 上海
项目面积 | 500 m²
主要材料 | 大理石、壁纸、布艺等

DESIGN CONCEPT | 设计理念

Positioned as modern and concise Chinese style, the primary principal of designs in this project is creating a modern Chinese space and presenting humanistic care to indoor environment according to local geographic environment and cultural atmosphere as well as paying attention to functions and practicability. Combining with train of thought of modern design, decorative designs in every floor make each area pursue unity in changes; the reasonable layout and remodeling of space create a primitive, elegant and comfortable interior space.

本案定位为现代简约中式风格，设计的首要原则是在注重功能，讲究实用的同时再根据当地的地理环境、人文气息，营造出具有现代中式风格的空间，体现出对人文关怀的室内环境。结合现代设计的思路，在每一层的装饰设计中，使区域在变化中追求统一，通过对空间的合理布局和重新塑造，创造一个古朴、典雅、舒适的室内空间。

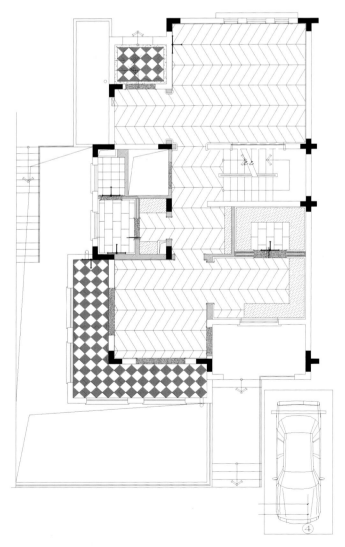

The tall living room is magnificent and full of quaint charm. The symmetrical hollow semicircle wall properly deduces the hide and exposure of Chinese style. The entire light elegant colors and floral wallpapers skillfully constitute a romantic dream and highlight the humanistic theme. In the villa, one can quietly express poetry.

挑高的客厅，大气而富有古韵。镂空对称的半圆造型墙，将中式的藏与露运用得恰到好处。而整体淡雅的色系，花纹的墙纸，巧妙地构筑起一个浪漫的梦，同时也更强调人文主题。身处别墅中，悄然话诗情！

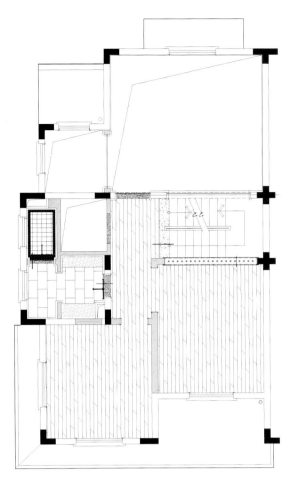
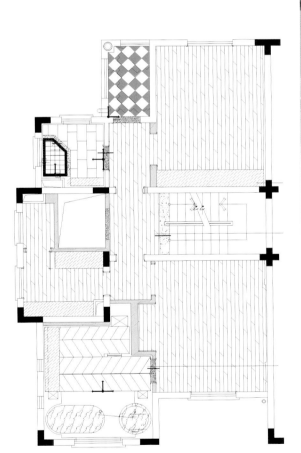

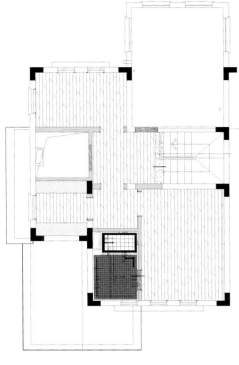
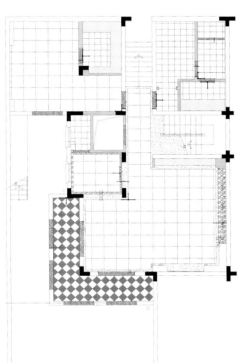

印迹 IMPRINT

杨星滨
XINGBIN YANG

临风听蝉　青墨雅涵
Listen to Cicadas with the Wind, Feel Cyan Ink and Elegant Connotation

设计公司｜一然设计
项目地点｜辽宁盘锦
项目面积｜444 m²
摄 影 师｜盛鹏
主要材料｜大理石、布艺、木质等

DESIGN CONCEPT ｜ 设计理念

This project focuses on the perfect combination of the essence of life and the beauty of art and the maximum realization of living function and artistic conception beauty.

The hard decoration adopts Oriental big-piece-cutting method; big opening and big closing deeply contain Oriental culture connotation; the big chopped piece is leisurely mastered and easily outlined, revealing the long history of Chinese culture in vivid images.

　　本案侧重于生活之本与艺术之美的完美结合，居住功能与意境美的最大化实现。

　　硬装上，采用东方大切块方式，大开大合，深含东方文化内蕴刀凿大形，斧切大块，从容把握，轻松勾勒，在生动的影像之中，透溢出来自华夏文化的渊远流长。

Sunshine enters from the tall space and large window, making the villa more luxurious and magnificent. The light elegant jade tone in the living room makes the space very elegant and graceful. The collocation of droplight and ruby chairs in the dining room and dining lamps inspired by bamboo light every candlelight dinner at home. The wood chairs in the study are charming. Every background and every image perfectly present Western techniques and Oriental charms. Everything is created by the designer to endow the owner's feelings into the entire space.

高挑的空间和偌大的开窗，阳光洒进室内，把别墅显得更加奢华气派。客厅淡雅的玉石色调将空间衬得格外优雅飘逸。餐厅吊灯与宝石色座椅的搭配，加上设计师由竹节引申出来的餐灯，点亮居家的每一场烛光晚餐。书房的木质桌椅行云流水烟波浩渺，一个背景，一个画面，都将西方技法与东方神韵完美呈现，一切都是设计师用笔将主人情怀无缝塑造于整个空间之中。

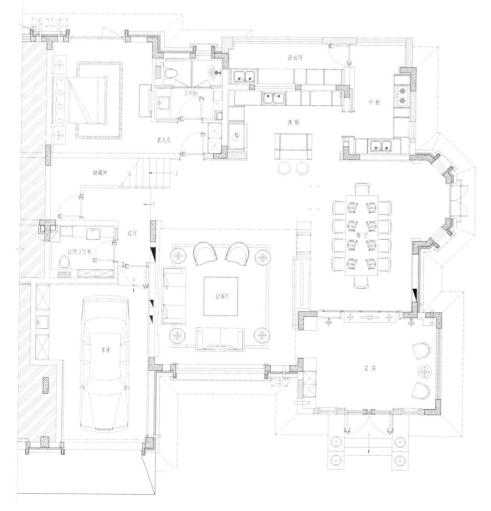

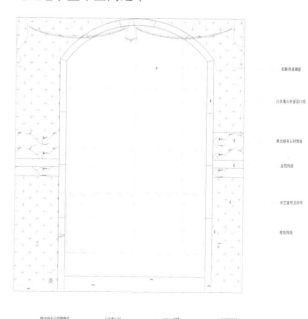

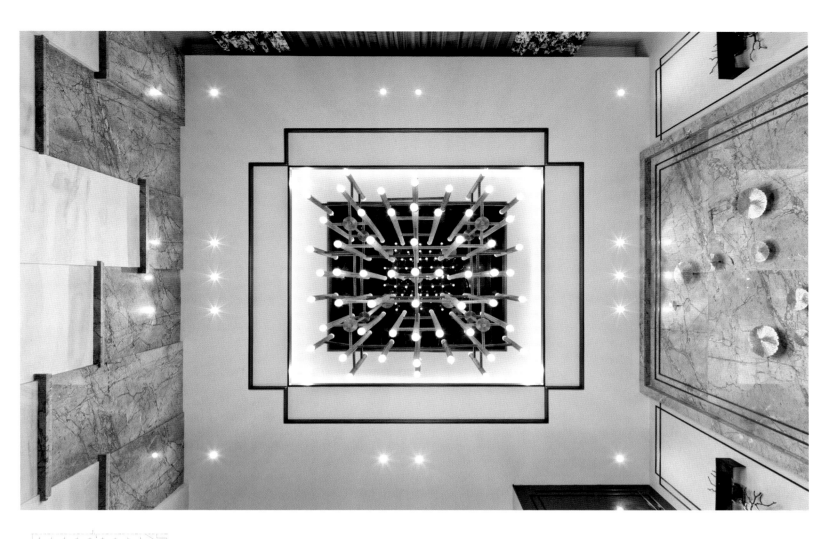

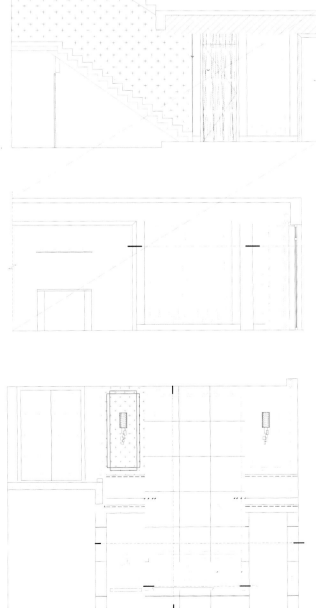

The soft decoration in the living room uses large area of gray ink as main melody, such as enjoyable peony, majestic landscape painting, crystal Taihu Lake stone and works of art. The decorative details outline the humanistic melody of modern Orient. The hall and dining room use real flame fireplace as the partition, which activates the space and embraces the world through the combination of Oriental and Western cultures. The purple and green in the bedroom create a new elegant and luxurious plot. The daughter's room boldly uses plum purple, gold and black, creating a traitorous space. Near the door, crane, lotus leaf and reed pond are the local natural states. The designer carefully integrates them to create a piece of beautiful landscape painting which is also the owner's hometown sentiments.

软装的陈设上，客厅大面积使用灰色水墨丹青的主旋律，写意的牡丹花，磅礴的山水画，水晶太湖石，艺术品等。在装饰细节上，勾勒出现代东方的人文主旋律。大厅和餐厅之间用真火壁炉做分隔，灵动了空间，结合东西方文化包罗世界。卧室的紫与绿，打造新派雅奢情节。女儿房采用大胆的梅红色、金色与黑色，呈现一个略带叛逆的空间。进门处仙鹤、荷叶、苇塘是当地的自然状态，设计师细心地把这些融合起来，打造一幅优美的山水画卷，也是主人的故乡情怀。

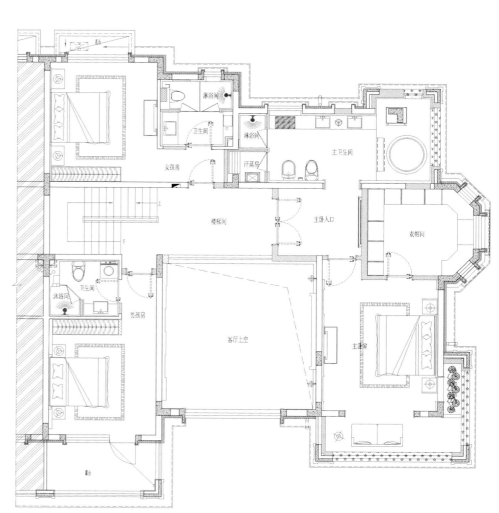

印迹 IMPRINT

宁睿
RUI NING

黛绿年华梦似锦
Splendid Dreams in Dark Green Times

项目名称 | 苏州工业园区海亮唐宁府
设计公司 | GND设计集团——N+恩嘉陈设
项目地点 | 江苏苏州
项目面积 | 160 m²
摄 影 师 | 大斌

DESIGN CONCEPT | 设计理念

This project adopts neo-Chinese style, which is not piles of Chinese symbols. It combines modern elements with traditional Oriental elements through understandings of traditional culture, reveals cultural connotations in solemn and elegant atmosphere, creates effects full of traditional flavors in modern people's aesthetic needs and perfectly presents traditional art in contemporary society. The plain and neat space flows quiet and meaningful tastes. Jumpy green in modern color scheme is the punchline to intersperse the space and also an important element to connect every space, which is harmonious and unified.

本作品采用新中式的风格进行演绎，不是中式符号的堆砌，而是通过对于传统文化的认识，将现代元素和传统东方元素结合在一起，从庄重典雅的氛围中流露文化底蕴，以现代人的审美需求来打造富有传统韵味的效果，让传统艺术在当今社会里得到完美体现。素净的空间流淌出宁静隽永的气息，现代配色手法以跳动的绿色作为点缀空间的亮笔之处，也是连接各空间的重要元素，使之浑然一体，和谐统一。

247

印迹 IMPRINT

陈赞 裴佳
ZAN CHEN
JIA PEI

幽深古色
Deep and Quiet Ancient Color

项目名称 ｜ 云南滇池高尔夫别墅
设计公司 ｜ 深圳市世纪方圆设计工程有限公司
摄 影 师 ｜ 张骑麟
主要材料 ｜ 木地板、玻璃、瓷砖等

DESIGN CONCEPT ｜ 设计理念

The tone of the entire space is profound and sedate with heavy Chinese meaning, mainly reflecting on the use of traditional furniture which is mainly from Ming and Qing Dynasties and the heavy and mature black furnishings. The interior design adopts symmetric layout, which is regular and moderate with elegant tone. Some traditional interior furnishings, such as calligraphy and painting, porcelains, antiques and antique-and-curio shelves, indicate the pursuit of self-cultivated life state. In this project, wood grilling works as a partition and a transparent furnishing, manifesting charms. Plain and elegant flowers bring delightful atmosphere; plain terrine and antique furnishing articles echo with the atmosphere of the space; the gentle and exquisite, elegant and Zen-like space fully convey the spirits of Chinese traditional aesthetics.

整体空间调性深沉稳重，中国风意味浓厚，主要表现在传统家具（多为明清家具）的使用及浓厚成熟的黑色系装饰色彩上。室内设计采用对称式的布局方式，端正稳健，格调高雅。通过一些传统的室内陈设如字画、陶瓷、古玩、博古架等，追求修身养性的生活境界。在本案例中，木质格栅栏，既起到分隔空间的作用，又是一种通透的装饰品，透出丝丝韵气。素色淡雅的花卉来增添愉悦的气氛，素胚的陶罐、古朴摆件与空间的氛围相呼应，平和雅致，儒雅禅意，充分传达出中国传统美学精神。

259

印迹
IMPRINT

杨铭斌
BENI YEUNG

泊墨之境
Ink State

项目名称 ｜ 誉海半岛9座101
设计公司 ｜ 硕瀚创研
软装设计 ｜ 东西无印
设计团队 ｜ 钟智豪、聂玉媚、潘嘉红
项目地点 ｜ 广东佛山
项目面积 ｜ 800 m²

DESIGN CONCEPT ｜ 设计理念

The designer hopes to use the neo-Chinese style of interior elements to match the tranquil outdoor garden. After entering the original building from the door, there is an independent entrance space with walls on both sides. The designer believes that in addition to dividing space, public space should also render the atmosphere, so the original walls on both sides are removed to put a semi-transparent screen with flowers and birds on it so as to introduce the garden scenery into indoors and complement with the artistic conception of the tea room.

设计师希望通过新中式室内元素的气韵与悠静的户外园林相映衬。原建筑从入户门进入室内是一个两边墙体的独立玄关空间，设计师认为公共空间之间应该在区分区域的基础上能够相互渲染，所以在设计上把两边原有的墙体拆除后放置半通透的花鸟元素屏风，以此把园林外的意境引入室内，同时与茶室的意境相得益彰。

Gentle and quiet tints are used as the main tone of the overall space, which is mainly made of wood veneer, white latex paint, wall cloth and titanium wire. The living room is 6-meter high, so the designer uses symmetric and concise lines at the living room through organizing the space to determine the axis lien and pursues details in the line-plate connection while processing the façade in terms of density ratio. Then, the originally not symmetric background walls are more harmonious and serious, demonstrating not only the atmosphere of the space, but also the attitudes to pursue details in life.

The semi-transparent screen plays an important function of partition and is also a scene in the aisle, making people in the space feel they are seemingly on the mountain and in the mist, while the sometimes twitter is even more enjoyable.

整体空间色调主要采用文雅静谧的浅色调，主要由木饰面、白色乳胶漆、墙布、钛金拉丝不锈钢等主材构成。有着6米高的客厅区域，设计师通过空间梳理，针对客厅部分运用对称和简练的线面构成手法，确立出空间的中轴线，并且在立面处理上力求在疏密比例的线面衔接中做到延伸的细节追求，使原本不对称的背景墙更加和谐庄重，体现的不仅是空间熏陶出来的意境，更是对生活细节追求的态度。

半通透屏风在空间中起着重要的隔断功能的同时更是过道中的一道风景线，让游走在空间中的人犹如身临高山迷雾中，室外时而传来鸟鸣更是写意。

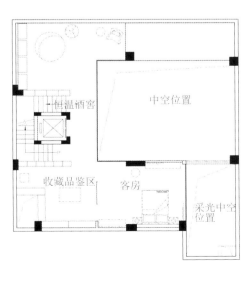

Except the height of the pipe layer, the negative floor one is approximately 4.5m high, an awkward height. Thus it will definitely be depressing to add an interlayer, so the designer puts the functional space where there are fewer movements in the interlayer while organizing the space function distributions and use the mirror and lighting effect to reduce the space's depressing sense to the minimum degree. The negative floor one also uses the line forms as floor one and storey height is reserved at the wine testing area to make the hollow position. Changes are made in materials by adding paint and titanium to make enrich colors in the space and make the space more elegant with the soft embellishment.

负一层去除管道层的高度约为4.5米，在如此尴尬的高度中做夹层固然会感到压抑，所以设计师在梳理空间功能分布的时候把走动较少的功能空间安置在夹层，并且通过镜面和灯光效果把空间压抑感减到最低。负一层同样延用一层的线面构成手法，品酒区保留层高做出中空位置。在材料上做变化，加入机理漆与钛金相搭配让空间色调更加丰富，在软装的点缀下更显高雅自得之意。

275

Reasonable space division and planning can improve the owners' life quality. The designer divides the entire three layers into activity area, and omits the original bath room to include it into the locker room, so a loop line is formed in the entire region.

Water and ink element on the background screen of the owner's room is the main vision of the space with mirror as the extension to mix the reality and virtual scene, adding fun and enriching the vision. The lamp groove around the ceiling is the rendering light source of the entire space just like the soft frog in the moonlight drifting throughout the space.

合理的空间规划能够提升业主的生活质量，设计师将整三层划分为主人活动区，并舍去原主卫生间的部分空间纳入到衣帽间，整个区域形成回路动线。

主人房床头背景的屏风水墨元素为空间的主视线，并以镜面延伸，虚实交错，增加了情趣，丰富了视觉。天花四周的灯槽是整个空间氛围的渲染光源，犹如在月光底下那水墨般的丝丝白雾，飘散贯穿于空间中。

IMPRINT 印迹

游滨绮
蔡曜牟
YU PIN-CHI
TSAI YAO-MOU

忆·相随 Memory and Accompanying

项目名称 ｜ 忆北京
设计公司 ｜ 创研侙集设计有限公司
项目地点 ｜ 台湾台北
项目面积 ｜ 363 m²
摄影师 ｜ 郭家和
主要材料 ｜ 木皮、ICI、蛇纹大理石、木地板等

DESIGN CONCEPT ｜ 设计理念

Used to work in Los Angeles and Beijing for more than 30 years, the capable and refined lady decides to return to her hometown after retirement. With more than 30 years reminiscences, the designers follow her past memories and new ideas to reinterpret them in this space and combine Oriental and Western elegant cultures to produce different design effects.

曾在洛杉矶及北京执业三十多年,干练且仪度娴雅的女士，退休后决定回到故乡。带着三十多年间的回忆拼图，依循着曾经的记忆及全新的想法，重新诠释于此空间，结合东西方优雅文化产生不同的设计效应。

With the entire red wall which is brought by paper-cutting in the public space and reflects walls of the Imperial Palace in the inner memory as the axis of the space, the overall space planning uses inner and outer relationships of corridor area and applies large area of French window with sufficient sunshine to bring outdoor scenery into interior. Natural colors intersperse a lot of Chinese antique furniture, creating a conflict and harmonious beauty, jumping out of the established design stereotypes and highlighting the integrating beauty of the design itself by using the layering and transformation of Orient and West.

整体的空间规划，以公共空间的老窗花带出的整面红墙（反映着记忆深处的故宫城墙）作为空间的轴心，使用廊道区划内、外间关系，运用阳光洒落的大片落地窗将窗外的植栽引景入室。自然中的色彩妆点空间大量中式古董家具放入空间设计，造就出冲突又协调的美感，并跳脱了既定的设计陈规，而取用东、西间的层次、转化，突出设计本身的融合之美。

IMPRINT 印迹

韩松
姚启盛

SONG HAN
QISHENG YAO

隐庐 Secluded Residence

项目名称 | 上海建发泗凤路Y1户型样板房
设计公司 | 深圳市昊泽空间设计有限公司
项目地点 | 上海
项目面积 | 120 m²
主要材料 | 石材、木地板、木饰面等

DESIGN CONCEPT | 设计理念

Dignified, elegant, refined and implicit charm owned by Chinese style is expressed by modern design languages. This kind of simple and clean space is more suitable for modern people to live in. Neo-Chinese living spaces are more and more popular in recent years, which reflects people's recognition to traditional living culture. The living room uses landscape painting as sofa background; light link traces slightly convey vacant artistic conception of landscape painting; white sofa matches with orange pillows; light color carpet and glass display tea table outline simple and elegant soul of the living room. The dining room chooses round dining table; black and wood dining table and chairs are concise and full of Chinese features of "simplifying complicated materials". The bedroom adopts orange and light color as main tones; the indoor green plants add a fresh vitality to it.

中式独有的端庄风华、清雅含蓄韵味以现代设计语言表达出来，这样素净的空间更适合现代人对居住的要求。近些年来新中式的居住空间愈发受欢迎，这也体现出人们对传统居住文化的认同。客厅以山水挂画为沙发背景，淡墨色的痕迹浅浅传递出山水写意画的空灵意境，白色沙发搭配橘色靠枕，浅色地毯、玻璃展示茶几勾勒出客厅素雅的精魂。餐厅中选择了圆形的餐桌，黑色和木色搭配的餐桌和餐椅，简约却充满中式"删繁去简"的特质。卧室以橘色和浅色为主色调，静置室内的绿植盆栽，为卧室增添一抹清新的生机。

印迹 IMPRINT

岳蒙
MENG YUE

一念姑苏 水云间
One Thought of Suzhou, Water and Cloud Residence

项目名称｜建邦·原香溪谷样板间
设计公司｜成象设计
软装公司｜成象软装
项目地点｜山东济南

DESIGN CONCEPT ｜ 设计理念

Trees and flowers are furnishings and the distant mountains are green; with a gentle breeze, the daylight and cloud shadows are wandering. Absorbing from spiritual waters, gaining brilliance from four seasons and leaving the space to nature are waiting for turns of seasons.

Using smoke and cloud as pen and ink and lying idle to enjoy the landscape, there is a piece of blue and a layer of blue; with heavy and light black and white and dizzy dark green, you can see fogs among mountains and smell moisture in the forest from the traditional Chinese realistic painting which imitates painting in Song Dynasty.

　　树花如缀，远山青黛含翠；清风徐来，天光云影徘徊。纳千钧之灵水，收四时之烂漫，把空间留给自然，是为了等待时节的流转。

　　烟云为笔墨，闲卧看山水，洒了一片青，又罩了一层蓝，黑白浓淡，青绿晕染，指下临摹宋画的工笔，似乎看得到山间雾岚，闻得到林间湿气。

Upon the table, there are mountains and rivers; near the pillows, there are bug buzz. Inviting a pair of green mountains into inside and a half pool of clean water to cook tea, the fun of mountain and forest doesn't lie between landscape which is far from hustle and bustle; if you have mountains and rivers inside, you can find a peach garden in busy downtown.

In the ancient atmosphere, a few pieces of exquisite stone are virtual, tranquil and calm. The white space can have unique connotation because of proper isolation and boundary. At first glance the chatters seem to be sounds of birds in forest. With pine, bamboo, wind and stove, inviting friends to share a pot of tea and a cup of tea, one may express the inner feelings when drunk. Flowers planted in four seasons are shaded by trees, making a natural idyllic scenery. With a stick of warm incense, time flies quickly like a dream; being in a peaceful and quiet place, everything can go with the wind. Rubbing an ink stick with the sound of wind, simple and elegant thing can ease the mind; with a piece of pen, the disposition can be released freely. The pine needles compete to show romance after a gentle breeze; conforming to nature can realize spiritual freedom and liberation. The light and shadow is leisure and tranquil pouring on the window, manifesting elegant charm of the ancient. At a spur of moment, simplicity and elegance appreciate each other; even in a corner, you can see the magnificent landscape. The environment is created by the heart; the furnishings are appropriate; with a piece of ink pen, the walls can be covered with floating clouds and flowing water.

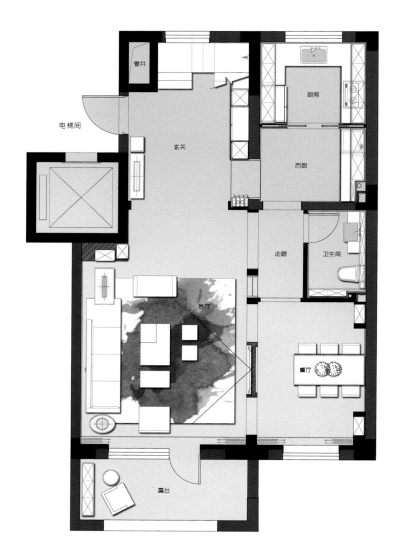

案上望山川，枕畔闻虫鸣，邀一对青山入座，请半潭清水烹茶，山林之趣，不在远离红尘的山水之间，胸中有丘壑，于闹市也可造一处桃花源。

时光若古，几块玲珑的石头，虚静恬然。隔而不隔，界而未界，留白处亦别有意蕴。乍听啁啁啾啾声，疑是林中鸟儿鸣。松风竹炉，提壶相呼，酒染诗情，醉时方吐胸中墨。四时插花，树痕掩映，自得天然画意。暖香一柱，恍若一梦，身在清净地，一切听风去。邀风声研墨，纳朴雅以逸志，只需一支笔，心性便可恣意挥洒。一叶松风竞烂漫，乘物以游心。光影悠幽倚窗纹，古人之雅兴尽显。

一时兴起，质朴与清雅惺惺相惜，只一角落，便可观山水大气象。境由心造，物与神游，借一支笔墨长篙，墙上便可满载行云流水。

印迹 IMPRINT

柏舍励创
PERCEPTRON DESIGN GROUP

端庄素影
Dignified and Plain Shadow

项目名称 | 佛山中海山湖世家
项目地点 | 广东佛山
项目面积 | 109 m²
摄 影 师 | 柏舍励创
主要材料 | 墙纸、硬包、瓷砖等

DESIGN CONCEPT | 设计理念

This project uses simple and practical designs and abandons visual impact of ritual senses to maintain Chinese concise texture in a long term.

The overall planning of the living room is given priority to lines. Roof lines and ground lines echo with each other. The main lamp chooses round lampshade, revealing Taoism of a square earth with a round sky above. The echoing of square and round makes the living room full of elegant and dignified senses. The color of TV wall is the main tone of the entire living room, which endows it with precipitated and tranquil flavors. The dining table and chairs in the dining room are classical, concise and fashionable. Chinese light and shade contrast lines integrate with modern art paintings. Lamp with wood texture has a flavor which is far away from the hustle and bustle. The Chinese long narrow table and hallstand in the guest room are placed with Chinese features. The whole room has a strong light and shade contrast. Wood floor, white wall and black lines add beauty to each other.

本案采用简约实用的设计摆放，摒弃了仪式感的视觉冲击，让中式简约质感久伴长远。

客厅整体规划以线条为主，房顶线条与地面线条相呼应，主灯选取圆形灯罩。透露着天圆地方的道家思想，而方与圆的呼应更让客厅整体充满端庄的感觉，影视墙的颜色选取是整个会客厅的主要色彩基调，让客厅显得更有岁月沉淀与宁静的味道。餐厅的餐桌座椅在古典简约中不失时尚，中式的明暗线条对比与现代艺术画相融合，木头质感的灯具有一种远离喧嚣的气息。客卧中式条案与衣帽架的摆放尽显中式特色。房间整体明暗对比强烈，木色地板、白色墙壁、黑色线条交相辉映。

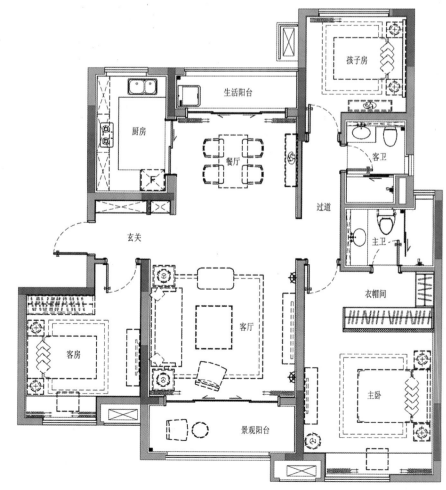

Details present Zen-like pursuits. The retro furniture and art works embody cultural connotations of the whole room.

细节之处彰显禅意追求，复古形态的家具、艺术品的摆放体现着房间整体的文化内涵。

印迹
IMPRINT

刘丽
LI LIU

繁华尽处 归于璞素
Simplicity After Gorgeous Prosperity

项目名称 ｜ 康桥九溪郡131叠拼样板间
设计公司 ｜ 筑详设计机构
参与设计 ｜ 宁静、孔令敏、张岩
项目地点 ｜ 广东东莞
项目面积 ｜ 325 m²
摄 影 师 ｜ 吴辉
主要材料 ｜ 米灰色亚麻壁布、老榆木洗白、拉斯古铜色不锈钢、雅士白石材、伊利诺灰石材等

DESIGN CONCEPT ｜ 设计理念

The vivid expression and strong sense of tableau in Chinese poetry appeal to people and embody the spirits and feelings of elegant people yearning for freedom, seclusion and nature. Currently, the noisiness and fickleness drive elite groups to pay a tribute to nature. We have an insight into the internal needs of customers but except the symbols of concretization, emphasize only the creation and expression of artistic atmosphere of the space to pursue a state of mind and attitude; and touch, experience and apperceive the essence of natural life by heart.

中国诗词的写意及强烈的画面感让人沉醉，体现贵族雅士向往自由、隐匿自然的精神情怀。当下的嘈杂与浮躁让精英群体集体开始向自然致敬，锁定客群的内在需求，不追求具象的符号，只关注空间意境氛围的营造与表达，追求一种心境，一份态度，用心聆听、体验、感悟自然生活的本质。

The first floor is a public activity space with ceremonial sense. The antecourt connects with the living room and dinning room. The kitchen is divided into two areas, the Chinese style and the western style. The Chinese style kitchen has two types, closed and open, connecting with the dinning area at the void area straightly; the western style kitchen and the bar counter form the shortest dynamic relation with the living room; the grey gallery connecting with the outside of big French windows can serve as a second living room with sunshine or a parlour to enjoy sunshine. It is an additional extension of the space. It is the best choice to enjoy four seasons with romantic wisteria. The soft decoration for hanging seat outdoors and occasional furniture accompanied by flowers and plants in the courtyard bring poetic quality to daily life; the room for the aged is private and independent. The home studio connects with the backyard. The owner may feel leisurely and carefree. The moso bamboo and the gurgling water form a quiet and refined courtyard. The dining table outdoors adds more delight of life in the courtyard. It is idyllic to invite several friends for free talks, painting or playing chess or piano, and reject all annoyances to rediscover the truest feelings.

一层为具有仪式感的公共活动空间，前庭连着客餐厅，中西厨被划分出两个区域，中厨设定封闭与开放两种模式,与挑空区的就餐区直线相连；西厨及吧台区与客厅的会客区形成最短的动线关系；与大落地窗的外部相连的灰空间结构长廊，可作为阳光次客厅或阳光休闲花厅使用，它是空间另外的延续，紫藤花下风花雪月的四季成为日常生活最好的背景，软装配置户外吊椅及休闲家具，搭配庭院花草，这一切让生活日常充满诗意；老人房私密独立，书房与后院相连，让生活悠闲自得，几促毛竹潺潺流水围合出静谧雅致的庭院，户外餐台为庭院增加更多的生活情趣。约上三五好友，谈天说地琴棋书画，让所有的烦恼与纷杂阻隔在精神之外，回归内心最真实的自己。

Two separate courtyards exist in the basement for host and hostess to develop hobbies independently and interactively. The natural lighting is sufficient in the space. The tranquil place enables customers with romantic feelings to enjoy staying alone and follow their hearts and experience the beauty of nature. The tasting area and storage area for red wine are arranged in the place free from the exposure to sunshine. The nurse room and laundry room are independent from the east line to the place where the host exercises in order to ensure the privacy and interactivity.

在地下一层开放出两个相对独立的院子，分隔出男女主人各自释放爱好的独立而又互动的空间，并确保空间的自然光照充足，让有情怀要求的客群享受独处及与内心对话并连接自然的静谧场所，把红酒品鉴及储藏区放在缺少阳光的暗区，另保姆间及其洗衣工作区与主人活动线分开，确保各个空间的私密性及互动性。

The second floor has living room and bed room, which are more private. The original architectural layout is modified to develop a space for the living room on the second floor. It may form visual and functional interaction with the living room on the first floor and provides more opportunities for parents to communicate with children. A room is cancelled to achieve the better results that the parents in the master bedroom and the kids in their rooms to have independent space and spiritual interaction.

The design with oriental expression not only meets the customers' needs but also embodies the appeal of designers who are found of the white space and artistic expression with oriental appeal and seek for design elements meeting spirits and daily life with depth and warm, after refining and recreation, it may produce new functions and visual tension and make the design return to the nature of space and the life. We care for behaviors of customers in daily life and the profundity of spiritual details to design and create more exquisite designs with a spirit of craftsman.

二层为更私密的起居室及卧室区域，通过对原有建筑布局的修改，在二楼开放出起居室的空间，既可与一楼客厅形成视觉及功能的互动，同时给家长与孩子的沟通提供更多机会。取消一个房间确保主卧及儿童房拥有更完善与独立的空间及精神关照。

东方意境的设计不光是对应客群的需求，同时也是设计师本人的诉求，喜欢留白、写意的有意境的东方情景的表达，寻求与精神契合并与生活日常相连的有深度有温度的设计元素，经过提炼再创造后让其行产生新的功能及视觉张力，让设计回归空间的本质，更回归生活的本质。关照客群日常的行为及精神细节的深入度，决定设计的深度，立志用匠人心态，打造更有深度的好设计。

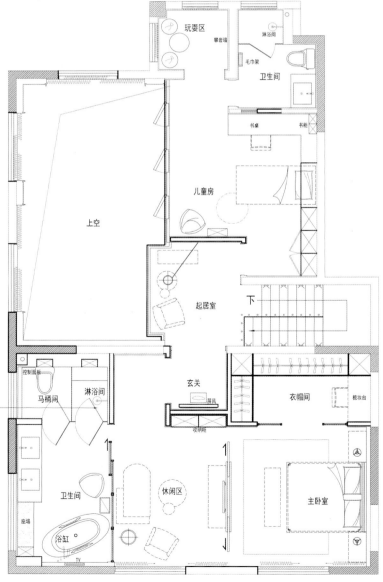

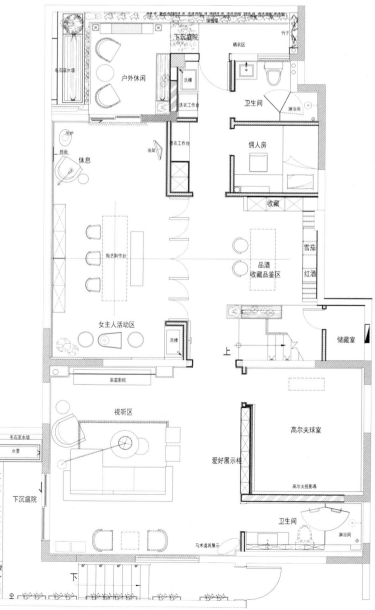

图书在版编目（CIP）数据

　　印迹：新中式居住空间 / 深圳视界文化传播有限公司编． -- 北京：中国林业出版社，2017.4
　　ISBN 978-7-5038-7895-4

　　Ⅰ．①印… Ⅱ．①深… Ⅲ．①住宅－室内装饰设计－图集 Ⅳ．① TU238.2-64

中国版本图书馆CIP数据核字（2017）第 055472 号

编委会成员名单
策划制作：深圳视界文化传播有限公司（www.dvip-sz.com）
总 策 划：万绍东
编　　辑：杨珍琼
装帧设计：潘如清
联系电话：0755-82834960

中国林业出版社 · 建筑分社
策　　划：纪　亮
责任编辑：纪　亮　王思源

出版：中国林业出版社
（100009 北京西城区德内大街刘海胡同 7 号）
http://lycb.forestry.gov.cn/
电话：（010）8314 3518
发行：中国林业出版社
印刷：深圳市雅仕达印务有限公司
版次：2017 年 5 月第 1 版
印次：2017 年 5 月第 1 次
开本：235mm×335mm，1/16
印张：20
字数：300 千字
定价：428.00 元（USD 86.00）